Synthesis Lectures on Ocean Systems Engineering

Series Editor

Nikolas Xiros, University of New Orleans, New Orleans, USA

The series publishes short books on state-of-the-art research and applications in related and interdependent areas of design, construction, maintenance and operation of marine vessels and structures as well as ocean and oceanic engineering.

Alexander Arnfinn Olsen

Means of Access for Enclosed Spaces

Guidance for Marine Professionals

Alexander Arnfinn Olsen
Southampton, UK

ISSN 2692-4420　　　　　　　　ISSN 2692-4471　(electronic)
Synthesis Lectures on Ocean Systems Engineering
ISBN 978-3-031-85865-9　　　　ISBN 978-3-031-85866-6　(eBook)
https://doi.org/10.1007/978-3-031-85866-6

© The Editor(s) (if applicable) and The Author(s), under exclusive license to Springer
Nature Switzerland AG 2026

This work is subject to copyright. All rights are solely and exclusively licensed by the Publisher, whether the whole or part of the material is concerned, specifically the rights of translation, reprinting, reuse of illustrations, recitation, broadcasting, reproduction on microfilms or in any other physical way, and transmission or information storage and retrieval, electronic adaptation, computer software, or by similar or dissimilar methodology now known or hereafter developed.
The use of general descriptive names, registered names, trademarks, service marks, etc. in this publication does not imply, even in the absence of a specific statement, that such names are exempt from the relevant protective laws and regulations and therefore free for general use.
The publisher, the authors and the editors are safe to assume that the advice and information in this book are believed to be true and accurate at the date of publication. Neither the publisher nor the authors or the editors give a warranty, expressed or implied, with respect to the material contained herein or for any errors or omissions that may have been made. The publisher remains neutral with regard to jurisdictional claims in published maps and institutional affiliations.

This Springer imprint is published by the registered company Springer Nature Switzerland AG
The registered company address is: Gewerbestrasse 11, 6330 Cham, Switzerland

If disposing of this product, please recycle the paper.

That's what a ship is, you know—it's not just a keel and a hull and a deck and sails, that's what a ship needs. But what a ship is, really is, is freedom.

—Captain Jack Sparrow, "Pirates of the Caribbean"

Preface

The maritime industry has long recognised that periodic surveys/inspections are the primary means for verifying that a vessel's structure is maintained within applicable requirements throughout its operational life. These surveys/inspections help confirm that the vessel is free from damage such as cracks, buckling, corrosion and overloading, and that material thickness are within established limits. For surveys/inspections to be carried out effectively, suitable means of access to the vessel's structure are required. To address the issue of suitable access, the Maritime Safety Committee (MSC) adopted the following resolutions into SOLAS Regulation II-1/3-6 on "Access to and within spaces in the cargo area of oil tankers and bulk carriers":

- MSC.194(80)—Adoption of Amendments to the International Convention for the Safety of Life at Sea, 1974, as amended,
- MSC.151(78)—Adoption of Amendments to the International Convention for the Safety of Life at Sea, 1974, as amended; and
- MSC.158(78)—Adoption of Amendments to the Technical Provisions.

To assist in the implementation of these new IMO resolutions, the International Association of Classification Societies (IACS) developed the Unified Interpretation (UI) SC 191, for the application of amended SOLAS Regulation II-1/3-6 (Resolution MSC.151(78)) and revised Technical Provisions for Means of Access for Inspections (Resolution MSC.158(78)). In addition, the Maritime Safety Committee (MSC) has approved the interpretations of the provision of SOLAS Chapter II-1 and of the Technical provisions adopted with resolution MSC.158(78) as contained in MSC.1/Circ.1284 and in MSC.1/Circ.1464/Rev.1.

Application

The vessel types and sizes listed below, if constructed on/after 1 January 2006 are required to comply with SOLAS Regulation II-1/3-6 and Resolutions MSC.151(78) and MSC.158(78):

- Oil tankers of 500 gross tonnage, and
- Bulk carriers (as defined in SOLAS regulation IX/1) of 20,000 gross tonnage and over.

Scope

In support of the Statutory and IACS guidance, this textbook (hereafter the "text") has been written to provide graphical illustrations and additional textual clarification of the means of access requirements. The goal is to improve the comprehension and application of the Statutory and IACS guidance.

Conceptually, this text presents two (2) levels of means of access guidance. The first provides the base criteria to meet the IMO requirements. The second, and preferred, level of guidance incorporates the application of current ergonomics practices to the means of access requirements. The application of ergonomics to the means of access requirements will enhance levels of safety as well as quality of work by providing appropriate access for survey, inspection and maintenance activities for tanks and holds.

Associated Notations

This text offers two optional notations related to means of access, **PMA** and **PMA +**. The **PMA** notation is awarded for compliance with all IMO permanent means of access requirements. The **PMA +** notation is awarded for compliance with the enhanced ergonomic level of means of access guidance contained in this text.

Southampton, UK　　　　　　　　　　　　　　　　　　　　Alexander Arnfinn Olsen

Acknowledgements

It is with grateful thanks that I acknowledge the following individuals for their support, insights and unwavering support in the development of this text:

- Dr. Kurt Moritz
- Dr. Werner Schmidt
- Dr. Franz Bertel

My sincere thanks and gratitude also go to Dr. Dieter Merkle, senior editor at Springer, and Boopalan Renu, editor at Straive, for supporting this project.

This title is published with the kind permission of the American Bureau of Shipping.

To you all, my deepest thanks and gratitude.

Alexander Arnfinn Olsen

Contents

1 **General Provisions** 1
   1.1 Introduction 1
   1.2 Applications 2
   1.3 Scope 2
   1.4 Common Terminology 3
   1.5 Notations 4
         PMA Notation 4
         PMA+Notation 4
   1.6 Detailed Considerations for the PMA+Notation 4
         Overview 4
         Example of the Application of the PMA+Notation 5
         Examples of PMA+Notation Opportunities 5
         IMO Means of Access Requirements Versus Ergonomic Practices 6
   1.7 Plans, Certification and Documentation 8
   1.8 Units of Measurement 8
   1.9 Alternatives 9
         General 9
         Other Regulations 9
         Departures from the Class Criteria 9

2 **Means of Access Requirements and Interpretations** 11
   2.1 General 11
   2.2 IMO Requirements for Means of Access 11
   2.3 Overview of the Means of Access Requirements 12
   2.4 Specific Means of Access Requirements for Oil Tankers 14
         Means of Access Requirements for Table 1/1.1 of MSC.158(78)
         for Oil Tankers (Tanks with a Height > 6 m (19.5 ft) and Containing
         Internal Structures) 14

Means of Access Requirements for Table 1/1.1.1 of MSC.158(78) for Oil Tankers (Tanks with a Height > 6 m (19.5 ft) and Containing Internal Structures)	16
Means of Access Requirements for Table 1/1.1.2 of MSC.158(78) for Oil Tankers (Tanks with a Height > 6 m (19.5 ft) and Containing Internal Structures)	17
Means of Access Requirements for Table 1/1.1.3 of MSC.158(78) for Oil Tankers (Tanks with a Height > 6 m (19.5 ft) and Containing Internal Structures)	18
Means of Access Requirements for Table 1/1.1.4 of MSC.158(78) for Oil Tankers (Tanks with a Height > 6 m (19.5 ft) and Containing Internal Structures)	19
Means of Access Requirements for Table 1/1.1.5 of MSC.158(78) for Oil Tankers (Tanks with a Height > 6 m (19.5 ft) and Containing Internal Structures)	20
Means of Access Requirements for Table 1/1.1.6 of MSC.158(78) for Oil Tankers (Tanks with a Height > 6 m (19.5ft) and Containing Internal Structures)	22
Means of Access Requirements for Table 1/1.2 of MSC.158(78) for Oil Tankers (Tanks with a Height < 6 m (19.5 ft))	23
Means of Access Requirements for Table 1/1.3 of MSC.158(78) for Oil Tankers (Fore Peak Tanks)	23
Means of Access Requirements for Table 1/1.3.1 of MSC.158(78) for Oil Tankers (Fore Peak Tanks)	24
Means of Access Requirements for Table 1/1.3.2 of MSC.158(78) for Oil Tankers (Fore Peak Tanks)	24
Means of Access Requirements for Table 1/2.1 of MSC.158(78) for Oil Tankers (Water Ballast Wing Tanks $\leq$5 m (16.5 ft) Width Forming Double Side Spaces and Their Bilge Hopper Sections)	24
Means of Access Requirements for Table 1/2.1.1 of MSC.158(78) for Oil Tankers (Double Side Spaces < 5 m (16.5 ft) in Width Above the Upper Knuckle Point of the Bilge Hopper Sections)	26
Means of Access Requirements for Table 1/2.1.2 of MSC.158(78) for Oil Tankers (Continuous Longitudinal Means of Access)	26
Means of Access Requirements for Table 1/2.1.3 of MSC.158(78) for Oil Tankers (Continuous Longitudinal Means of Access)	27
Means of Access Requirements for Table 1/2.2 of MSC.158(78) for Oil Tankers (Access for Distances $\leq$6 m (19.5 ft) from the Tank Bottom to the Upper Knuckle Point)	28
Means of Access Requirements for Table 1/2.2.1 of MSC.158(78) for Oil Tankers (Continuous Longitudinal Means of Access)	30

	Means of Access Requirements for Table 1/2.2.2 of MSC.158(78) for Oil Tankers (Continuous Longitudinal Means of Access)	30
	Means of Access Requirements for Table 1/2.3 of MSC.158(78) for Oil Tankers (Access for Distances < 6 m (19.5 ft) from the Tank Bottom to the Upper Knuckle Point)	32
2.5	Specific Means of Access Requirements for Bulk Carriers	33
	Means of Access Requirement for Table 2/1.1 of MSC.158(78) for Bulk Carriers (Cargo Holds—Access to Underdeck Structure)	33
	Means of Access Requirement for Table 2/1.2 of MSC.158(78) for Bulk Carriers (Cargo Holds—Access to Underdeck Structures)	35
	Means of Access Requirement for Table 2/1.3 of MSC.158(78) for Bulk Carriers (Cargo Holds—Access to Underdeck Structure)	36
	Means of Access Requirement for Table 2/1.4 of MSC.158(78) for Bulk Carriers (Cargo Holds—Access to Underdeck Structure)	36
	Means of Access Requirement for Table 2/1.5 of MSC.158(78) for Bulk Carriers (Cargo Holds—Access to Underdeck Structure)	37
	Means of Access Requirement for Table 2/1.6 of MSC.158(78) for Bulk Carriers (Cargo Holds—Access to Vertical Structures)	38
	Means of Access Requirement for Table 2/1.7 of MSC.158(78) for Bulk Carriers (Cargo Holds—Access to Vertical Structures)	38
	Means of Access Requirement for Table 2/1.8 of MSC.158(78) for Bulk Carriers (Cargo Holds—Access to Vertical Structures)	39
	Means of Access Requirement for Table 2/1.9 of MSC.158(78) for Bulk Carriers (Cargo Holds—Access to Vertical Structures)	40
	Means of Access Requirement for Table 2/1.10 of MSC.158(78) for Bulk Carriers (Cargo Holds—Access to Vertical Structures)	40
	Means of Access Requirement for Table 2/1.11 of MSC.158(78) for Bulk Carriers (Cargo Holds—Access to Vertical Structures)	41
	Means of Access Requirement for Table 2/2.1 of MSC.158(78) for Bulk Carriers (Ballast Tanks—Top Side Tanks)	41
	Means of Access Requirement for Table 2/2.2 of MSC.158(78) for Bulk Carriers (Ballast Tanks—Top Side Tanks)	42
	Means of Access Requirement for Table 2/2.3 of MSC.158(78) for Bulk Carriers (Ballast Tanks—Top Side Tanks)	44
	Means of Access Requirement for Table 2/2.4 of MSC.158(78) for Bulk Carriers (Ballast Tanks—Top Side Tanks)	44
	Means of Access Requirement for Table 2/2.5 of MSC.158(78) for Bulk Carriers (Ballast Tanks—Bilge Hopper Tanks)	45
	Means of Access Requirement for Table 2/2.5.1 of MSC.158(78) for Bulk Carriers (Ballast Tanks—Bilge Hopper Tanks)	46

	Means of Access Requirement for Table 2/2.5.2 of MSC.158(78) for Bulk Carriers (Ballast Tanks—Bilge Hopper Tanks)	47
	Means of Access Requirement for Table 2/2.5.3 of MSC.158(78) for Bulk Carriers (Ballast Tanks—Bilge Hopper Tanks)	48
	Means of Access Requirement for Table 2/2.6 of MSC.158(78) for Bulk Carriers (Ballast Tanks—Bilge Hopper Tanks)	49
	Means of Access Requirement for Table 2/2.7 of MSC.158(78) for Bulk Carriers (Ballast Tanks—Bilge Hopper Tanks)	50
	Means of Access Requirement for Table 2/2.8 of MSC.158(78) for Bulk Carriers (Ballast Tanks—Bilge Hopper Tanks—Double-Skin Side Tanks)	50
	Means of Access Requirement for Table 2/2.9 of MSC.158(78) for Bulk Carriers (Ballast Tanks—Fore Peak Tanks)	51
	Means of Access Requirement for Table 2/2.9.1 of MSC.158(78) for Bulk Carriers (Ballast Tanks—Fore Peak Tanks)	51
	Means of Access Requirement for Table 2/2.9.2 of MSC.158(78) for Bulk Carriers (Ballast Tanks—Fore Peak Tanks)	52

3 Walkways, Ramps and Work Platforms 53
   3.1 General 53
          Design Loads 53
   3.2 Walkways and Ramps 54
          General Principles 54
          Toe Boards 55
          Walkway and Ramp Design 57
   3.3 Work Platforms 57
          General Principles 58

4 Vertical Ladders, Inclined Ladders and Handles 61
   4.1 General 61
          Design Loads 61
          Use and Selection of Ladders 62
   4.2 Vertical Ladders 63
          General Principles 63
          Vertical Ladder Design 63
          Climber Safety Devices 65
          Fall Protection from Secondary Fall Points 70
          Individual Rung Ladders 74
   4.3 Inclined Ladders 76
          General 78
          Inclined Ladder Design 78
          Spiral Ladders 79

	4.4	Handles	79
		General	82
		Handle Design/placement	83
5	**Hatches**		87
	5.1	General	87
	5.2	General Principles	87
	5.3	Hatch Design	88
	5.4	Horizontal Hatch Access Near a Coaming	88
	5.5	Horizontal Hatch Access Near a Coaming	89
		Horizontal Hatch Access Through a Deck (for PMA+notation)	90
6	**Alternative Means of Access**		93
	6.1	General	93
		Definitions	93
	6.2	Application of SOLAS and MSC Regulations	94
	6.3	IMO Requirements for Alternative Means of Access	94
	6.4	Guidance for Alternative Means of Access	95
		Portable Ladders	95
		Hydraulic Arm Vehicles	97
		Wire Lift Platform	98
		Portable Platforms	99
		Scaffolding and Staging	100
		Rafting	101
		Remotely Operated Vehicle (ROV)	104
Appendix			105

Acronyms and Abbreviations

°	Degrees
ASTM	American Society of Testing and Materials
ft	Feet
IACS	International Association of Classification Societies
IMO	International Maritime Organisation
in	Inch
m	Metre
mm	Millimetre
MSC	Maritime Safety Committee
PMA	Class notation signifying that the vessel's permanent means of access meets IMO and IACS requirements
PMA+	Class notation signifying that the vessel's permanent means of access meets IMO and IACS requirements plus additional ergonomic considerations
SOLAS	Safety of Life at Sea

List of Figures

Fig. 2.1	Access at transverse bulkhead on stiffened side of an underdeck structure for ballast/cargo tanks ≥6 m (19.5 ft) in height	16
Fig. 2.2	Continuous longitudinal access on each side of the tank of the underdeck structure for ballast/cargo tanks ≥6 m (19.5 ft) in height	18
Fig. 2.3	Integrated continuous longitudinal access of a longitudinal bulkhead of cargo tanks or holds	20
Fig. 2.4	Access for cross-ties ≥6 m (19.5 ft) above the tank bottom of cargo tanks or holds	21
Fig. 2.5	Access for cross-ties ≥6 m (19.5 ft) above the tank bottom of cargo tanks or holds—alternative arrangement	22
Fig. 2.6	Access where the vertical distance between the horizontal uppermost stringer and deck hold is ≥6 m (19.5 ft)	28
Fig. 2.7	Access for bilge hopper sections where the vertical distance from the tank bottom to the upper knuckle point is ≥6 m (19.5 ft)	29
Fig. 2.8	Access for bilge hopper tank	31
Fig. 2.9	Alternative means of access	32
Fig. 2.10	Access for vertical distances <6 m (19.5 ft) from the tank bottom to the upper knuckle point	34
Fig. 2.11	Access to underdeck structures ≥17 m (56.0 ft) in height	35
Fig. 2.12	Athwartship access fitted on the transverse bulkhead ≥17 m (56.0 ft) in height	36
Fig. 2.13	Bulk carriers having transverse bulkheads with full upper stools ≥17 m (56.0 ft) in height	39
Fig. 2.14	Upper topside tank access with a height >6 m (19.5 ft)	42
Fig. 2.15	Transverse web access for bulk carriers	43
Fig. 2.16	Bilge hopper tank access with a height ≥6 m (19.5 ft)	45

Fig. 2.17	Access ladder between the longitudinal continuous means of access and the bottom of the space	46
Fig. 2.18	Alternate bilge hopper tank access with a height ≥ 6 m (19.5 ft)	48
Fig. 2.19	Access for foremost and aftmost bilge hopper tanks	49
Fig. 3.1	Toe board dimensions	55
Fig. 3.2	Walkway and ramp design	56
Fig. 3.3	Web frame walkways	57
Fig. 3.4	Discontinuous handrail where top and mid rails are connected[+]	58
Fig. 3.5	Discontinuous handrails where top and mid rails are not connected[+]	59
Fig. 3.6	Work platform dimensions	60
Fig. 4.1	Vertical ladders (general criteria)	64
Fig. 4.2	Staggered vertical ladder **a** side mount **b** ladder through the linking platform	66
Fig. 4.3	Vertical ladders to landings (side mount)[+]	69
Fig. 4.4	Vertical ladders to landings (ladder through platform)[+]	70
Fig. 4.5	Arrangement for cage of vertical ladder	71
Fig. 4.6	Cage of vertical ladder—side view	72
Fig. 4.7	Ladders with climber safety rails or cables	73
Fig. 4.8	Front view of guardrail requirements for vertical ladders without safety cages or climber safety rails/cables	75
Fig. 4.9	Side view of guardrail requirements for vertical ladders without safety cages or climber safety rails/cables	76
Fig. 4.10	Front view of guardrail requirements for vertical ladders with safety Cages and without climber safety rails/cables*	77
Fig. 4.11	Individual rung ladder design	79
Fig. 4.12	Inclined ladders	80
Fig. 4.13	Inclined ladders with landings	81
Fig. 4.14	Inclined ladder landing/platform	82
Fig. 4.15	Handle dimensions	83
Fig. 4.16	Handle placement (ladder not extending through platform)[+]	84
Fig. 4.17	Handle placement (stepping through a vertical hatch)[+]	85
Fig. 4.18	Handle placement (stepping to or from a vertical ladder)	86
Fig. 5.1	Hatch design	88
Fig. 5.2	Hatch design (alternative arrangement)	89
Fig. 5.3	Handle placement (stepping through a vertical hatch)	90
Fig. 5.4	Access hatch heights of ≥ 900 mm (35.5 in)	91
Fig. 5.5	Horizontal hatch access through a deck	92

List of Tables

Table 2.1	Application of resolution MSC.158(78) Table 1 for oil tankers	12
Table 2.2	Application of resolution MSC.158(78) Table 2 for bulk carriers	13
Table 4.1	Selection of access type	63
Table 4.2	Guardrail requirements for vertical ladders without safety cages or climber safety rails/cables	74
Table 4.3	Guardrail requirements for vertical ladders with safety cages and without climber safety rails/cables	78

General Provisions

1.1 Introduction

The ability to survey/inspect the condition of a vessel is a principal means to help verify that the vessel's structure is maintained to comply with applicable requirements. These surveys/inspections assist asset owners, Flag State Administrations and classification societies in determining that vessels are free from damage and that material thickness is within established limits. For surveys/inspections to be carried out safely and effectively, suitable means of access to the vessel's structure is required. The most recent adoptions to IMO legislation (SOLAS Reg. II-1/3-6) have established new requirements for means of access. Throughout the life of a vessel, this access enables overall and close-up inspections and material thickness measurements of the vessel's structures. These means of access may be used by Flag State Administrations, classification societies, vessel personnel and others, as necessary.

To assist in the implementation of these new requirements, IACS has developed Unified Interpretation (UI) SC 191 for the application of amended SOLAS regulation II-1/3-6 (resolution MSC.151 (78)) and revised Technical provisions for means of access for inspections (resolution MSC.158 (78)).In addition, the Maritime Safety Committee (MSC) has approved the interpretations of the provision of SOLAS Chapter II-1 and of the Technical provisions adopted with resolution MSC.158(78) as contained in MSC.1/Circ.1284 and in MSC.1/Circ.1464/Rev.1. In support of this document, this text has been developed. This text seeks to provides additional information, via text and graphics, about the means of access requirement's interpretation and application, as well as the criteria for the **PMA** and **PMA+** (or alternative Class equivalent) notations.

1.2 Applications

The means of access requirements in SOLAS and in this text apply to:

- Oil tankers of 500 gross tonnage and over constructed on or after 1 January 2006. This regulation is only applicable to oil tankers having tanks integral with the structure of the vessel which are used for carriage of oil in bulk, which is contained in the definition of oil in Annex 1 of MARPOL 73/78. Independent oil tanks can be excluded; and/or
- Bulk carriers (as defined in SOLAS regulation IX/1) of 20,000 gross tonnage and over, constructed on or after 1 January 2006. SOLAS Regulation IX/1 defines a bulk carrier as a ship which are generally constructed with single deck, topside tanks and hopper side tanks in cargo spaces, and is intended primarily to carry dry cargo in bulk and includes such types as ore carriers and combination carriers.

Note: *Oil tankers of 500 gross tonnage and over constructed on or after 1 October 1994 but before 1 January 2005 should comply with the provisions of regulation II-1/12-2 adopted by resolution MSC.27(61). Also, for oil tankers of less than 5,000 tonnes deadweight, Flag State Administrations may approve, in special circumstances, smaller dimensions for access through vertical and horizontal openings, if the ability to traverse such openings or to remove an injured person can be proved to the satisfaction of the Flag State Administration.*

1.3 Scope

To enable physical surveys, inspections, and maintenance activities to be conducted effectively, consideration needs to be given to how the vessel will be designed, in particular, means of access arrangements. The effectiveness of the design of means of access can be maximised, as illustrated in this text, through the application of ergonomics to the vessel's structural designs and arrangements. This text presents two (2) levels of means of access guidance. The first level provides the base criteria to meet the IMO requirements. The second, and preferred, level of guidance incorporates the application of ergonomics to the IMO means of access requirements. This text also presents a generic process to obtain Class approval for the use of alternative materials, equivalent in strength and stiffness to steel, for the construction of means of access as allowed for in MSC.151 (78) and MSC.158 (78).

This text overlaps, in several areas, with the text *Applying Physical Ergonomics to Modern Ship Design* (Olsen; Springer: 2024). This text is a recommended companion to further promote the application and understanding of ergonomic principles to vessel designs. The application of ergonomics to the means of access requirements can improve overall personnel performance and safety, while reducing the potential for human error.

1.4 Common Terminology

Accessibility: The ability for personnel to access equipment that requires maintenance, inspection, removal, or replacement while wearing the appropriate clothing, including personal protective equipment, and the ability to use all necessary tools and test equipment.

Active protection: A safety design or device that actively (or directly) requires a person to take specific actions before a potential loss, for example, donning a fall arrestor fitted to both the ladder and the climber.

Crew member: Any person onboard a vessel, including the Master, who is not a passenger. This term is used interchangeably throughout this text with "seafarer."

Design load: The maximum intended load, being the total of all loads including the weight of the personnel, materials, and equipment, including the means of access structure.

Guardrail or safety rail: Device for protection against accidental fall or accidental access to a hazardous area, with which stairs, step ladders or landings, platforms and walkways, or deck edges/fall points may be equipped.

Handrail: Top element designed to be grasped by the hand for body support which can be used individually or as the upper part of a rail.

Maintenance: All activities necessary to keep equipment in, or restore it to, a specified level of performance.

Newton: SI unit of force. One Newton is equal to the amount of net force required to accelerate a mass of one kilogram at a rate of one metre per second squared.

Newton-metres (N-m): SI unit of torque. One Newton-metre is equal to the torque resulting from a force of one Newton applied perpendicularly to a moment arm which is one metre long.

Seafarer: Any person onboard a vessel, including the Master, who is not a passenger. This term is used interchangeably throughout this document with "crew member."

Should: Expresses a provision that is mandatory.

Toe board: Solid lower part of a guard-rail on a landing to prevent the fall of objects from a floor level.

1.5 Notations

This text offers two optional notations related to means of access, **PMA** and **PMA+**. The **PMA** notation is awarded for compliance with all IMO permanent means of access requirements. The **PMA+** notation is awarded for compliance to the enhanced ergonomic level of means of access guidance contained in this text.

PMA Notation

The optional **PMA** notation may be assigned to vessels to signify that the vessel's permanent means of access meets the following:

- IMO Resolution MSC.151(78)—"Adoption of amendments to the international convention for the safety of life at sea, 1974, as amended,"
- IMO Resolution MSC.158(78)—"Adoption of amendments to the technical provisions for means of access for inspections," and
- IACS Unified Interpretation (UI) SC 191 for the application of amended SOLAS regulation II-1/3-6 (resolution MSC.151 (78)) and revised Technical provisions for means of access for inspections (resolution MSC.158 (78))

Note: The **PMA** *notation does not include IMO requirements for alternative or temporary means of access.*

PMA+ Notation

The optional **PMA+** notation may be assigned to vessels meeting the **PMA** notation requirements plus the additional ergonomic considerations presented in this text. The **PMA+** notation is discussed in more detail in the following section.

1.6 Detailed Considerations for the PMA+ Notation

Overview

The **PMA+** notation applies ergonomics and safety design practices to the design and arrangement of the permanent means of access requirements where allowable. These instances include those areas where the means of access requirements have prescribed minimums and/or maximums and ergonomic dimension exists within the allowable range.

1.6 Detailed Considerations for the PMA+ Notation

For those means of access requirements where no specific dimensioning is provided, ergonomic guidance and dimensioning is provided.

Note: If any **PMA+** *requirements are found to conflict with MSC requirements or IACS interpretations, the MSC requirements or IACS interpretations take precedence.*

Example of the Application of the PMA+ Notation

For tankers, IMO states "Continuous athwartship permanent access arranged at each transverse bulkhead on the stiffened surface, at a minimum of 1.6 m to a maximum of 3 m below the deck head." This 1.6 m (63.0 in) minimum is below current ergonomic practices, which is a minimum of 2130 mm (84.0 in). The minimum overhead clearance of 1600 mm (63.0 in) is approximately 117 mm (4.5 in) less than the mean male stature (height). Also, this mean value does not include clothing allowances of approximately 25 mm (1.0 in) for footwear and 75 mm (3.0 in) for safety helmets. By combining these two clothing allowances to the mean stature we have an average male height of 1817 mm (71.5 in). This value is approximately 220 mm (8.5 in) higher than the IMO minimum height.

For a large majority of personnel this 1600 mm (63.0 in) minimum would increase the likelihood of head strikes against overhead surfaces, potentially causing head and neck injuries as well as causing personnel to work/walk in awkward postures causing potential upper and lower back pain, discomfort or injury.

The current ergonomic practice of a 2130 mm (84.0 in) minimum overhead clearance will allow the vast majority of personnel to work upright without the potential for head strikes against upper surfaces or from working in awkward postures.

The way IMO states the overhead clearances, from "a minimum of 1.6 m to a maximum of 3 m" provides designers the opportunity to use the **PMA+** requirement of 2130 mm (84.0 in) as a design requirement.

Examples of PMA+ Notation Opportunities

The majority of the dimensional aspects of the means of access requirements are stated in a manner that provides the designer with some latitude with respect to dimensioning. Several examples include:

- "The minimum clear opening should not be less than 600 mm × 600 mm"—which means that the clear opening can be greater than the dimension specified,
- "Elevated passageways forming sections of a permanent means of access, where fitted, should have a minimum clear width of 600 mm"—which means the passageway can have a greater clear width,

- "Stanchions should be not more than 3 m apart"—this establishes a maximum distance only. A shorter dimension is allowed,
- "Permanent inclined ladders should be inclined at an angle of less than 70°"—which means that inclined ladders cannot exceed 70°, but inclined ladders can be at less of an angle. Additionally, there are instances where no dimensional aspects of the means of access requirements are provided; and
- "Inclined ladders should be provided with handrails of substantial construction on both sides fitted at a convenient distance above the treads"—no dimensional requirements are provided for the size of the handrail or the handrail's height above the tread or any intermediate rails.

IMO Means of Access Requirements Versus Ergonomic Practices

There are other instances of IMO means of access requirements, which could be enhanced through the application of ergonomic practices but cannot because the specific wording of the IMO means of access requirements prohibits any modification of these requirements. These instances include the design of guardrail heights, openings in horizontal stringers, and the tread design and spacing of inclined ladders. These are discussed in more detail in the following paragraphs.

Note: The discussion in the following paragraphs (sections "Guardrail Heights", "Tread Spacing (Inclined Ladders)" and "Tread Design (Inclined Ladders)") are not **PMA+** *requirements. These paragraphs discuss examples of where ergonomics practices should be applied but cannot be due to the specific wording of the IMO means of access requirements.*

Guardrail Heights

MSC.158 (78), Technical Provisions 3.3, requirement states that "… guardrails shall be 1000 mm (39.5 in) in height and consist of a rail and an intermediate bar 500 mm in height and of substantial construction…" This specific wording allows no opportunity to apply ergonomic principles to the design of guardrails. From an ergonomics and safety perspective a guardrail's height should be 1070 mm (approximately 42.0 in). The rational for this change in guardrail height is based on the biomechanics (including centre of gravity) of a worker leaning or falling over a guardrail. Biomechanical analyses show that the centre of gravity of a human body is approximately 75 mm (3.0 in) above the midpoint of a person's stature (the centre of gravity varies slightly among individuals and genders). Biomechanical analyses also show that if the centre of gravity of a human body act above a guardrail, a person falling against the rail would have a higher tendency to rotate over the top of the railing. Additionally, if the centre of gravity of a human body act below the top of a rail, a person would tend to rotate under the railing. This action (rotating under) along with intermediate rail(s) will help prevent a fall to a lower surface.

1.6 Detailed Considerations for the PMA+ Notation

To illustrate this point, a person 1830 mm (72.0 in) tall would have a centre of gravity of approximately 1000 mm (39.5 in). This means that people with a stature greater than 1830 mm (72.0 in), including footwear, would have a higher likelihood of rotating over the top of a 1000 mm (39.5 in) guardrail than a 1070 mm (42.0 in) guardrail. The requirement for a guardrail height of at least 1000 mm (39.5 in) is more effective for personnel less than 1830 mm (72.0 in) tall (including footwear), leaving those workers who are taller at a safety disadvantage. Current ergonomics design practices for the height of guardrails take into consideration taller potential workers (up to 2130 mm (84.0 in) in height). When considering the taller potential worker population, a guardrail height of 1070 mm (42.0 in) will help protect approximately 99% of all workers.

Tread Spacing (Inclined Ladders)
MSC.158 (78), Technical Provision 3.6, "… the treads shall be equally spaced at a distance apart, measured vertically, of between 200 and 300 mm…" From an ergonomic perspective, the range provided for the spacing of treads should be expanded to 180–300 mm (7.0–12.0 in). It is noted that there is not much difference between the MSC requirements and the current ergonomic practice for the tread spacing for inclined ladders. The reasoning behind the current ergonomic practice is based on designing for the 5th percentile (shorter) female and to provide a wider range and opportunity to make sure the inclined ladder treads are equally spaced throughout the flight of the ladder.

Tread Design (Inclined Ladders)
MSC.158 (78), Technical Provisions 3.6, "… when steel is used, the treads shall be formed of two square bars of not less than 22 mm × 22 mm in section, fitted to form a horizontal step with the edges pointing upward…" Walking on the corner edges of square bars provides for less contact with the foot and the stepping surface and an increased opportunity for slipping. Also, standing on the corner edges for extended period of time (e.g., performing an inspection) causes pressure points and pain on the bottom of the foot. The preferred tread design should be a solid plate or circular cross section bars. Treads should be formed of a solid step (not less than 100 mm (4.0 in) in depth) with non-slip surface or non-slip circular cross section bars (not less than 25 mm (1.0 in) in diameter). If the steps are constructed using circular cross-section bars (which is preferred so as to minimise sludge accumulations on treads of ladders fitted in cargo oil tanks), they should consist of two or more parallel bars arranged on the same horizontal plane, with the distance between the centres of adjacent bars being not less than 65 mm (2.5 in) and not more than 75 mm (3.0 in). The IMO does not provide any guidance with respect to step depth or the spacing of the adjacent bars.

1.7 Plans, Certification and Documentation

One set of the official (design and construction agents) electronic copies of the following plans and information should be submitted to the Class engineering department for the purpose of review in the context of the notation being sought:

(1) Details of arrangements of the components and structures appropriate for the notation(s) being sought,
(2) Diagrammed details of each of the above components and structures; and
(3) Any vendor documentation or certifications pertinent to applying the requirements to the design.

For new construction, the drawings should be provided to the Class engineering department during the detailed design phase. For existing vessels, the arrangement drawings and plans reflecting the current configurations (e.g., topsides, below decks, etc.) should be provided to, and approved by, the Class engineering department in advance of Class surveyor verifications. The Class engineering department should review the submitted accommodations documentation. The Class engineering department Engineering should report any deviation from criteria to the asset owner/shipyard for resolution and should also identify any criteria that the Class surveyors must field-verify. The Class surveyor should verify that the submitted drawings match the constructed vessel. The Class surveyor should also verify any criteria that are outstanding from the Class engineering department review and document deviations from criteria.

1.8 Units of Measurement

This text uses both SI and US units of measure. Within this document the SI unit is listed first with the US unit of measure in parentheses. The SI unit indicates the driving measure, not the US unit. The US units are provided for the convenience of general users. To maintain consistency between other ergonomic and human factors related documentation, the same US unit of measure rounding scheme has been used in this document. US units of measure expressed in inches (in) are rounded to the nearest half inch (0.5 in). US units of measure expressed in feet (ft) are rounded to the one-half foot (0.5 ft).

1.9 Alternatives

General

Class will consider alternative arrangements or criteria which can be shown to meet the intent of criteria directly cited or referred to in this text. The demonstration of an alternative's acceptability can be made through either the presentation of satisfactory service experience or systematic analysis based on valid engineering principles.

Other Regulations

Class may consider for its acceptance, alternative arrangements and details which can be shown to comply with standards recognised in the country (Flag State) in which the vessel is registered or operated, provided they are deemed not less effective.

Departures from the Class Criteria

It is recognised that unusual or unforeseen conditions may lead to a case where one or more of the parameters of interest in granting a notation may temporarily fall outside the range of acceptability. When a departure from criteria is identified during either the notation's initial issuance or reconfirmation process, it should be reviewed by Class in consultation with the asset owner. When the ergonomic design contains departures from the stated criteria, these will be subject to special consideration upon the receipt of details about the departure. Depending on the degree and consequences of the departure, the shipyard or asset owner may be required to provide an assessment and remediation plan to obtain or maintain the notation. Failure to complete the agreed remediation by the due date will lead to withdrawal of the notation.

Means of Access Requirements and Interpretations

2.1 General

This chapter presents the means of access requirements as set forth in MSC.151(78), MSC Resolution MSC.158(78), and the associated IACS Unified Interpretations. Chapter 2, section "Overview of Means of Access Requirements" presents a summary of the IMO requirements in tabular form. Chapter 2, section "Specific Means of Access Requirements for Oil Tankers" and Chap. 2, section "Specific Means of Access Requirements for Bulk Carriers" present the actual requirements and interpretations contained in the MSC and the IACS documentation. Where appropriate, additional guidance, via text and graphics, has been provided to further promote the interpretation and application of the means of access requirements.

2.2 IMO Requirements for Means of Access

The following documents provide specific details about the IMO requirements for means of access contained in this section:

- SOLAS regulation II-1/3-6, "Access to and within spaces in, and forward of, the cargo area of oil tankers and bulk carriers,"
- IMO Resolution MSC.151 (78) (adopted on 20 May 2004), "Adoption of amendments to the international convention for the safety of life at sea, 1974, as amended,"
- IMO Resolution MSC.158 (78) (adopted 20 May 2004), "Amendments to the technical provisions for means of access for inspections;" and

- IACS Unified Interpretation (UI) SC 191 for the application of amended SOLAS regulation II-1/3-6 (resolution MSC.151 (78)) and revised Technical provisions for means of access for inspections (resolution MSC.158 (78)).

2.3 Overview of the Means of Access Requirements

This section summarises the means of access requirements for oil tankers and bulk carriers as presented in MSC.158(78). The following two (2) tables, Table 2.1 "Application of resolution MSC.158(78) Table 1 for oil tankers" and Table 2.2, "Application of resolution MSC.158(78) Table 2 for bulk carriers" were created to simplify the use and application of the extensive tables containing the MSC means of access requirements. These tables are arranged with column and row headings for different areas of vessel tanks and holds/spaces as well as tank or hold/space dimensions (heights and widths). This format allows for the quick identification of means of access requirements. To reduce potential confusion for users of this text who may already be familiar with the MSC.158(78) tables, the requirements contained in this text's Table 2.1 summarise the requirements contained in the MSC.158(78) Table 1. This text's Table 2.2 summarises the requirements contained in the MSC.158(78) Table 2.

Table 2.1 Application of resolution MSC.158(78) Table 1 for oil tankers*

Cargo/ballast tank	Tank height ≥ 6 m (19.5 ft)	Tank height < 6 m (19.5 ft		
Underdeck structure	1.1.1, 1.1.2, 1.1.3	1.2		
Longitudinal bulkhead	1.1.4 or 1.1.6$^+$			
Cross tie (≥6 m (19.5 ft) above tank bottom)	1.1.5			
Ballast tank and double side skin space	Tank/Space width ≥ 5 m (16.5 ft)	Tank/Space width < 5 m (16.5 ft)		
		Height ≥ 6 m (19.5 ft)	Height < 6 m (19.5 ft)	
Wall-sided mid-depth portion (between topside and hopper portions)	1.1.4 or 1.1.6$^+$	2.1	Not applicable	
Lower hopper portion/tank	1.1.5	2.2	2.3	
Fore peak tanks	1.3	1.3		

* = Numbers in this table correspond to MSC.158 (78) Table 1, "Means of access for ballast/cargo tanks of oil tankers"
+ = If height < 17 m (56.0 ft)

2.3 Overview of the Means of Access Requirements

Table 2.2 Application of resolution MSC.158(78) Table 2 for bulk carriers[*]

Cargo/ballast tank	Tank height ≥ 6 m (19.5 ft)	Tank height < 6 m (19.5 ft)	
Underdeck structure	1.1, 1.2, 1.3, 1.4	1.5	
Side shell	1.6[+] or 1.8, 1.7, 1.9[+], 1.10[+]	Not applicable	
Vertical bulkhead	1.7		
Ballast tank and double side skin space	Tank/Space width ≥ 5 m (16.5 ft)	Tank/Space width < 5 m (16.5 ft)	
		Height ≥ 6 m (19.5 ft)	Height < 6 m (19.5 ft)
Wall-sided mid-depth portion (between topside and hopper portions)	2.8	2.8	
Upper topside tanks	2.1, 2.2, 2.3	2.4	
Lower hopper portion/tank	2.5 and 2.6	2.7	
Fore peak tanks	2.9	2.9	

[*] = Numbers in this table correspond to MSC.158 (78) Table 2, "Means of access for bulk carriers.")
[+] = Single side skin construction only

To further simplify the use of these tables, the numbering scheme used in the MSC tables has been preserved. For example, in Table 2.1, in the Underdeck Structure row, tanks with a height of 6 m or more are required to meet MSC.158(78) Table 1 requirements of 1.1.1, 1.1.2, and 1.1.3. The same rational has been applied to MSC.158(78) Table 2.

2.4 Specific Means of Access Requirements for Oil Tankers

The specific means of access requirements for oil tankers displayed in Table 2.1 "Application of resolution MSC.158(78) Table 1 for oil tankers", as interpreted by IACS UI SC 191, are presented in this chapter.

Note: Those requirements that relate to permanent means of access are prerequisites for the **PMA** notation. The **PMA** notation does not include the IMO requirements or allowances for alternative or temporary means of access.

Each MSC requirement is presented in its entirety along with any associated IACS unified interpretation. Graphical representations for some of the means of access requirements have been provided to help clarify or demonstrate the requirement's intent. Also, additional guidance related to the design of the means of access is provided.

Note: Graphics are for demonstrative purposes only and are not to scale.

In some of the figures, specific dimensions are provided. These dimensions will help clarify the means of access requirement and/or the interpretation. Where appropriate, IMO, **PMA** and **PMA+** dimensional requirements are provided. These dimensions are to be used in conjunction with the additional guidance contained in the following sections:

- Chapter 3, "Walkways, Ramps and Work Platforms",
- Chapter 4, "Vertical Ladders, Inclined Ladders and Handles", and
- Chapter 5, "Hatches."

Means of Access Requirements for Table 1/1.1 of MSC.158(78) for Oil Tankers (Tanks with a Height >6 m (19.5 ft) and Containing Internal Structures)

Means of access requirements for Table 1/1.1 of MSC.158(78) for oil tankers		
PMA requirement	Means of access requirement	"For tanks of which the height is 6 m and over containing internal structures, permanent means of access shall be provided in accordance with .1 to .6:"

(continued)

2.4 Specific Means of Access Requirements for Oil Tankers

(continued)

Means of access requirements for Table 1/1.1 of MSC.158(78) for oil tankers

	IACS interpretation	• Sub-paragraphs .1, .2 and .3 define access to underdeck structure, access to the uppermost sections of transverse webs and connection between these structures • Sub-paragraphs .4, .5 and .6 define access to vertical structures only and are linked to the presence of transverse webs on longitudinal bulkheads • If there are no underdeck structures (deck longitudinals and deck transverses) but there are vertical structures in the cargo tank supporting transverse and longitudinal bulkheads, access in accordance with sub-paragraphs from .1 through to .6 is to be provided for inspection of the upper parts of vertical structure on transverse and longitudinal bulkheads • If there is no structure in the cargo tank, section 1.1 of Table 1 is not to be applied • Section 1 of Table 1 is also to be applied to void spaces in cargo area, comparable in volume to spaces covered by the regulation II-1/3-6, except those spaces covered by Section 2 • The vertical distance below the overhead structure is to be measured from the underside of the main deck plating to the top of the platform of the means of access at a given location • The height of the tank is to be measured at each tank • For a tank the height of which varies at different bays, item 1.1 is to be applied to such bays of a tank that have height 6 m (19.5 ft) and over
PMA+ requirement		No additional **PMA+** requirements
Additional guidance		• Detailed walkway design guidance is available in Chap. 3, "Walkways, Ramps and Work Platforms" • Detailed ladder and handle guidance is available in Chap. 4, "Vertical Ladders, Inclined Ladders and Handles" • Detailed guidance for hatch design is available in Chap. 5, "Hatches"

Means of Access Requirements for Table 1/1.1.1 of MSC.158(78) for Oil Tankers (Tanks with a Height >6 m (19.5 ft) and Containing Internal Structures)

Means of access requirements for Table 1/1.1 of MSC.158(78) for oil tankers				
PMA requirement	Means of access requirement	"Continuous athwartship permanent access arranged at each transverse bulkhead on the stiffened surface, at a minimum of 1.6 m to a maximum of 3 m below the deck head"		
		Dimension		MSC.158(78) requirement
		A	Distance below deckhead	$\geq$1600 mm (63.0 in) and $\leq$3 m (10.0 ft)
	IACS interpretation	"The vertical distance below the overhead structure is to be measured from the underside of the main deck plating to the top of the platform of the means of access at a given location"		
PMA+ requirement		Dimension		**PMA+** requirement
		A	Distance below deckhead	$\geq$2130 mm (84.0 in) and $\leq$3 m (10.0 ft)
Additional guidance		• Figure 2.1, "Access at transverse bulkhead on stiffened side of an underdeck structure for ballast/cargo tanks $\geq$6 m in height" is provided to help illustrate this requirement • Detailed walkway design guidance is available in Chap. 3, section "Walkways and Ramps" • Continuous athwartship **PMA** is usually arranged at the same level as integrated structural members (horizontal girders)		

Fig. 2.1 Access at transverse bulkhead on stiffened side of an underdeck structure for ballast/cargo tanks $\geq$6 m (19.5 ft) in height

2.4 Specific Means of Access Requirements for Oil Tankers

Means of Access Requirements for Table 1/1.1.2 of MSC.158(78) for Oil Tankers (Tanks with a Height >6 m (19.5 ft) and Containing Internal Structures)

Means of access requirements for Table 1/1.2 of MSC.158(78) for oil tankers

PMA requirement	Means of access requirement	"At least one continuous longitudinal permanent means of access at each side of the tank. One of these accesses shall be at a minimum of 1.6 m to a maximum of 6 m below the deck head and the other shall be at a minimum of 1.6 m to a maximum of 3 m below the deck head"		
		Dimension		MSC.158(78) requirement
		A	Distance below deckhead	≥1600 mm (63.0 in) and ≤3 m (10.0 ft)
		B	Distance below deckhead	≥1600 mm (63.0 in) and ≤6 m (19.5 ft)
	IACS interpretation	"The vertical distance below the overhead structure is to be measured from the underside of the main deck plating to the top of the platform of the means of access at a given location"		
PMA+ requirement		Dimension		**PMA+** requirement
		A	Distance below deckhead	≥2130 mm (84.0 in) and ≤3 m (10.0 ft)
		B	Distance below deckhead	≥2130 mm (84.0 in) and ≤6 m (19.5 ft)
Additional guidance		• Figure 2.2, "Continuous longitudinal access on each side of the tank of the underdeck structure for ballast/cargo tanks ≥6 m in height" is provided to help illustrate these requirements • Distances "A" and "B" below the deckhead are measured from the underside of the plating to the top of the platform • Detailed walkway design guidance is available in Chap. 3, section "Walkways and Ramps" • Detailed hatch design guidance is available in Chap. 5, "Hatches"		

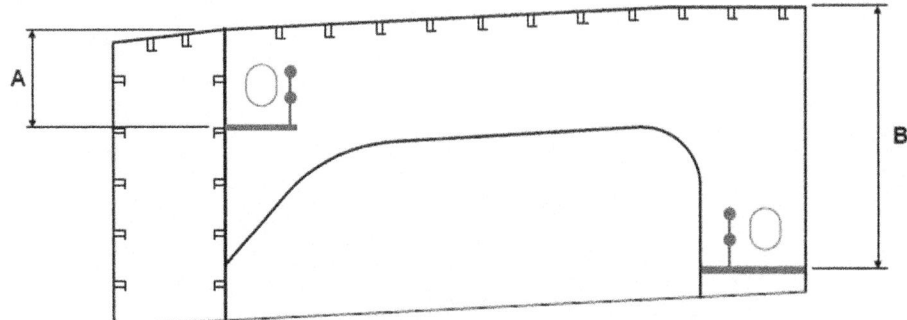

Fig. 2.2 Continuous longitudinal access on each side of the tank of the underdeck structure for ballast/cargo tanks ≥6 m (19.5 ft) in height

Means of Access Requirements for Table 1/1.1.3 of MSC.158(78) for Oil Tankers (Tanks with a Height >6 m (19.5 ft) and Containing Internal Structures)

Means of access requirements for Table 1/1.3 of MSC.158(78) for oil tankers		
PMA requirement	Means of access requirement	"Access between the arrangements specified in 1.1.1 and 1.1.2 and from the main deck to either 1.1.1 or 1.1.2"
	IACS interpretation	"Means of access to tanks may be used for access to the permanent means of access for inspection"
PMA+ requirement		No additional **PMA+** requirements
Additional guidance		• Detailed walkway design guidance is available in Chap. 3, "Walkways, Ramps and Work Platforms" • Detailed guidance for hatch design is available in Chap. 5, "Hatches"

2.4 Specific Means of Access Requirements for Oil Tankers

Means of Access Requirements for Table 1/1.1.4 of MSC.158(78) for Oil Tankers (Tanks with a Height >6 m (19.5 ft) and Containing Internal Structures)

Means of access requirements for Table 1/1.4 of MSC.158(78) for oil tankers

PMA requirement	Means of access requirement	"Continuous longitudinal permanent means of access which are integrated in the structural member on the stiffened surface of a longitudinal bulkhead, in alignment, where possible, with horizontal girders of transverse bulkheads are to be provided for access to the transverse webs unless permanent fittings are installed at the uppermost platform for use of alternative means, as defined in paragraph 3.9 of the Technical provisions, for inspection at intermediate heights"
	IACS interpretation	"The permanent fittings required to serve alternative means of access such as wire lift platform, that are to be used by crew and surveyors for inspection shall provide at least an equal level of safety as the permanent means of access stated by the same paragraph. These means of access shall be carried on board the ship and be readily available for use without filling of water in the tank. Therefore, rafting is not acceptable under this provision" • "Alternative means of access are to be part of Access Manual, which is to be approved on behalf of the Flag State" • For water ballast tanks of 5 m or more in width, such as on an ore carrier, side shell plating shall be considered in the same way as "longitudinal bulkhead"
PMA+ requirement		No additional **PMA+** requirements
Additional guidance		• Figure 2.3, "Integrated continuous longitudinal access of a longitudinal bulkhead of cargo tanks or holds", shows two views of this requirement • Graphic "A" shows the intent of this requirement • Graphic "B" shows an alternative design if permanent fittings (e.g., to accommodate a wire lift platform) are installed at the uppermost platform to obtain access to the vertical web • Detailed walkway design guidance is available in Chap. 3, section "Walkways and Ramps"

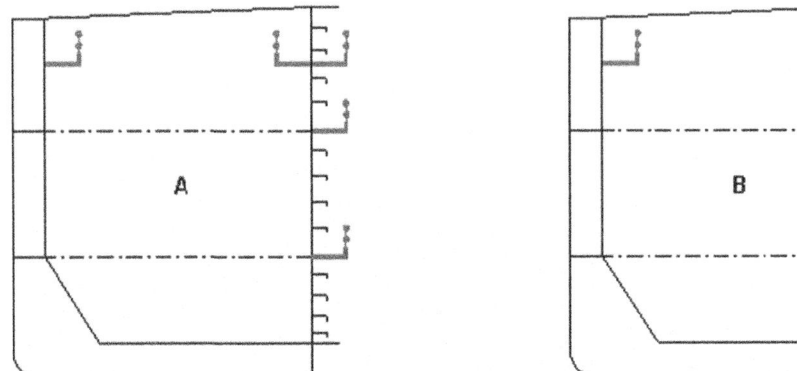

Fig. 2.3 Integrated continuous longitudinal access of a longitudinal bulkhead of cargo tanks or holds

Means of Access Requirements for Table 1/1.1.5 of MSC.158(78) for Oil Tankers (Tanks with a Height >6 m (19.5 ft) and Containing Internal Structures)

Means of access requirements for Table 1/1.5 of MSC.158(78) for oil tankers				
PMA requirement	Means of access requirement	"For ships having cross-ties which are 6 m or more above tank bottom, a transverse permanent means of access on the cross-ties providing inspection of the tie flaring brackets at both sides of the tank, with access from one of the longitudinal permanent means of access in 1.1.4"		
		Dimension		MSC.158(78) requirement
		A	Handrail height	≥1000 mm (39.5 in)
	IACS interpretation	No interpretation provided		
PMA+ requirement		A continuous walkway (as illustrated below) is provided for at least 25% of all cross-ties in a tank, or at least one (1) per tank, whichever is greater		

(continued)

2.4 Specific Means of Access Requirements for Oil Tankers

(continued)

Means of access requirements for Table 1/1.5 of MSC.158(78) for oil tankers

Additional guidance	• Figure 2.4, "Access for cross-ties ≥6 m above the tank bottom of cargo tanks or holds" is provided to help illustrate this requirement. This figure shows a continuous walkway extending across the cross-tie. This ergonomic improvement will help enhance inspection and facilitate quicker access to flaring brackets on both sides of the tank • Figure 2.5, "Access for cross-ties ≥6 m above the tank bottom of cargo tanks or holds—alternative arrangement" is provided to help illustrate an alternative solution when longitudinal permanent means of access is provided at each end of the cross tie • An important design consideration is the distance of the walkway below the top of the cross-tie. The distance of the walkway below the cross-tie shall be approximately 1000 mm (39.5 in). This will help eliminate a potential falling hazard • Detailed walkway design guidance is available in Chap. 3, section "Walkways and Ramps" • Detailed hatch design guidance is available in Chap. 5, "Hatches." Distances "A" and "B" below the deckhead are measured from the underside of the plating to the top of the platform • Detailed walkway design guidance is available in Chap. 3, section "Walkways and Ramps" • Detailed hatch design guidance is available in Chap. 5, "Hatches"

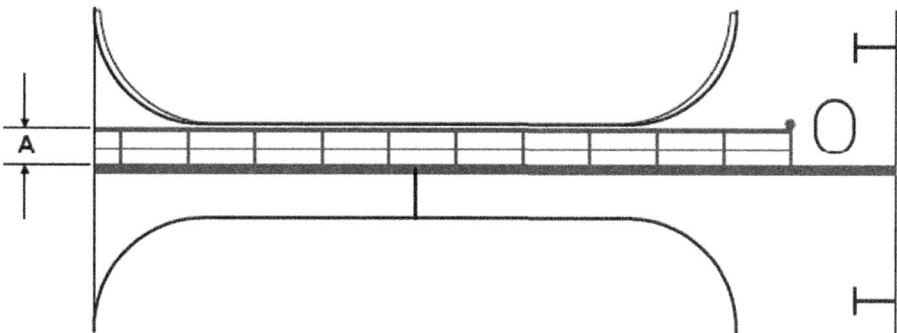

Fig. 2.4 Access for cross-ties ≥6 m (19.5 ft) above the tank bottom of cargo tanks or holds

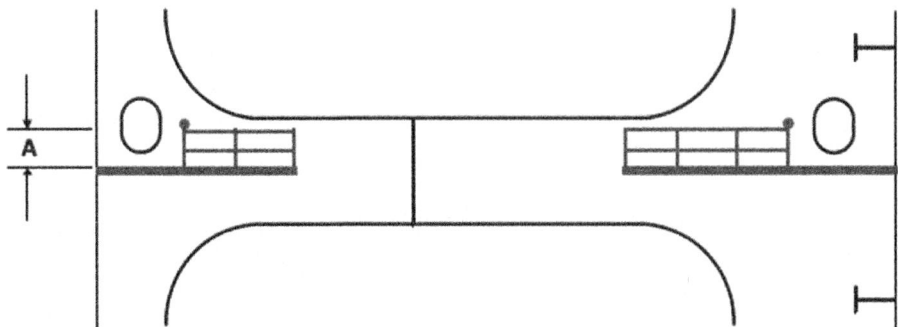

Fig. 2.5 Access for cross-ties ≥6 m (19.5 ft) above the tank bottom of cargo tanks or holds—alternative arrangement

Means of Access Requirements for Table 1/1.1.6 of MSC.158(78) for Oil Tankers (Tanks with a Height >6 m (19.5ft) and Containing Internal Structures)

Means of access requirements for Table 1/1.6 of MSC.158(78) for oil tankers		
PMA requirement	Means of access requirement	"Alternative means as defined paragraph 3.9 in the Technical provisions may be provided for small ships as an alternative to 1.1.4 for cargo oil tanks of which the height is less than 17 m"
	IACS interpretation	No interpretation provided
PMA+ requirement		No additional **PMA+** requirements
Additional guidance		Additional guidance on the design and selection of alternative means of access is available in Chap. 6, "Alternative means of access"

Means of Access Requirements for Table 1/1.2 of MSC.158(78) for Oil Tankers (Tanks with a Height <6 m (19.5 ft))

Means of access requirements for Table 1/1.2 of MSC.158(78) for oil tankers		
PMA requirement	Means of access requirement	"For tanks of which the height is less than 6 m, alternative means as defined in paragraph 3.9 of the Technical provisions or portable means may be utilised in lieu of the permanent means of access"
	IACS interpretation	No interpretation provided
PMA+ requirement		No additional **PMA+** requirements
Additional guidance		• Additional guidance on the design and selection of alternative means of access is available in Chap. 6, "Alternative means of access"

Means of Access Requirements for Table 1/1.3 of MSC.158(78) for Oil Tankers (Fore Peak Tanks)

Means of access requirements for Table 1/1.3 of MSC.158(78) for oil tankers		
PMA requirement	Means of access requirement	"Stringers of less than 6 m in vertical distance from the deck head or a stringer immediately above are considered to provide suitable access in combination with portable means of access"
	IACS interpretation	No interpretation provided
PMA+ requirement		No additional **PMA+** requirements
Additional guidance		• Detailed walkway design guidance is available in Chap. 3, section "Walkways and Ramps" • Additional guidance on the design and selection of alternative means of access is available in Chap. 6, "Alternative Means of Access"

Means of Access Requirements for Table 1/1.3.1 of MSC.158(78) for Oil Tankers (Fore Peak Tanks)

Means of access requirements for Table 1/1.3.1 of MSC.158(78) for oil tankers

PMA requirement	Means of access requirement	"In case the vertical distance between the deck head and stringers, stringers or the lowest stringer and the tank bottom is 6 m or more, alternative means of access as defined in paragraph 3.9 of the Technical provisions shall be provided"
	IACS interpretation	No interpretation provided
PMA+ requirement		No additional **PMA+** requirements
Additional guidance		• Detailed walkway design guidance is available in Chap. 3, section "Walkways and Ramps" • Additional guidance on the design and selection of alternative means of access is available in Chap. 6, "Alternative Means of Access"

Means of Access Requirements for Table 1/1.3.2 of MSC.158(78) for Oil Tankers (Fore Peak Tanks)

Means of access requirements for Table 1/1.3.2 of MSC.158(78) for oil tankers

PMA requirement	Means of access requirement	"In case the vertical distance between the deck head and stringers, stringers or the lowest stringer and the tank bottom is 6 m or more, alternative means of access as defined in paragraph 3.9 of the Technical provisions shall be provided"
	IACS interpretation	No interpretation provided
PMA+ requirement		No additional **PMA+** requirements
Additional guidance		• Detailed walkway design guidance is available in Chap. 3, section "Walkways and Ramps" • Additional guidance on the design and selection of alternative means of access is available in Chap. 6, "Alternative Means of Access"

Means of Access Requirements for Table 1/2.1 of MSC.158(78) for Oil Tankers (Water Ballast Wing Tanks ≤5 m (16.5 ft) Width Forming Double Side Spaces and Their Bilge Hopper Sections)

2.4 Specific Means of Access Requirements for Oil Tankers

Means of access requirements for Table 1/1.3.2 of MSC.158(78) for oil tankers

PMA requirement	Means of access requirement	"For double side spaces above the upper knuckle point of the bilge hopper sections, permanent means of access are to be provided in accordance with .1 to .3;"
	IACS interpretation	"Section 2 of Table 1 (The above requirement—4.10 MSC Requirement 1.3.2 (Fore Peak Tanks) is also to be applied to wing tanks designed as void spaces"
PMA+ requirement		No additional **PMA+** requirements
Additional guidance		• Detailed walkway design guidance is available in Chap. 3, section "Walkways and Ramps" • Detailed hatch design guidance is available in Chap. 5, "Hatches"

Means of Access Requirements for Table 1/2.1.1 of MSC.158(78) for Oil Tankers (Double Side Spaces <5 m (16.5 ft) in Width Above the Upper Knuckle Point of the Bilge Hopper Sections)

Means of access requirements for Table 1/2.1.1 of MSC.158(78) for oil tankers

PMA requirement	Means of access requirement	"Where the vertical distance between horizontal uppermost stringer and deck head is 6 m or more, one continuous permanent means of access shall be provided for the full length of the tank with a means to allow passing through transverse webs installed a minimum of 1.6 m to a maximum of 3 m below the deck head with a vertical access ladder at each end of tank"	
		Dimension	MSC.158(78) requirement
		A — Distance below deckhead	≥1600 mm (63.0 in) and ≤3 m (10.0 ft)
	IACS interpretation	• "This paragraph (2.1.1) represents requirements for access to underdeck structures • For a tank the vertical distance between horizontal upper stringer and deck head of which varies at different sections item 2.1.1 is to be applied to such sections that fall under the criteria • The continuous permanent means of access may be a wide longitudinal, which provides access to critical details on the opposite side by means of platforms as necessary on web frames. In case the vertical opening of the web frame is located in way of the open part between the wide longitudinal and the longitudinal on the opposite side, platforms shall be provided on both sides of the web frames to allow safe passage through the web frame • Where two access hatches are required by SOLAS regulation II-1/3-6.3.2, access ladders at each end of the tank are to lead to the deck"	
PMA+ requirement		Dimension	PMA+ requirement
		A — Distance below deckhead	≥2130 mm (84.0 in) and ≤3 m (10.0 ft)
Additional guidance		• Figure 2.6, "Access where the vertical distance between the horizontal uppermost stringer and deck hold is ≥6 m" is provided to help illustrate these requirements • Distance "A" below the deckhead is measured from the underside of the plating to the top of the platform • Detailed walkway design guidance is available in Chap. 3, section "Walkways and Ramps" • Detailed hatch design guidance is available in Chap. 5, "Hatches"	

Means of Access Requirements for Table 1/2.1.2 of MSC.158(78) for Oil Tankers (Continuous Longitudinal Means of Access)

Means of access requirements for Table 1/2.1.2 of MSC.158(78) for oil tankers

PMA requirement	Means of access requirement	Continuous longitudinal permanent means of access, which are integrated in the structure, at a vertical distance not exceeding 6 m (19.5 ft) apart	
		Dimension	MSC.158(78) requirement
		A — Access height	≤6 m (19.5 ft)

(continued)

2.4 Specific Means of Access Requirements for Oil Tankers

(continued)

Means of access requirements for Table 1/2.1.2 of MSC.158(78) for oil tankers

	IACS interpretation	• "This paragraph (2.1.2) is a requirement for access for survey and inspection of vertical structures on longitudinal bulkheads (transverse webs) • The continuous permanent means of access may be a wide longitudinal, which provides access to critical details on the opposite side by means of platforms as necessary on web frames. In case the vertical opening of the web is located in way of the open part between the wide longitudinal and the longitudinal on the opposite side, platforms shall be provided on both sides of the web to allow safe passage through the web. A 'reasonable deviation' as noted in TP/1.4, of not more than 10% may be applied where the permanent means of access is integral with the structure itself"
PMA+ requirement		No additional **PMA+** requirements
Additional guidance		• Detailed walkway design guidance is available in Chap. 3, section "Walkways and Ramps" and Chap. 3, section "Work Platforms" • Detailed hatch design guidance is available in Chap. 5, "Hatches" • With the Administration's approval, "reasonable deviations" may be applied to facilitate this means of access. IACS UI (SC) 191 has interpreted this to be no more than 10% for vertical distances exceeding 6 m (19.5 ft)

Means of Access Requirements for Table 1/2.1.3 of MSC.158(78) for Oil Tankers (Continuous Longitudinal Means of Access)

Means of access requirements for Table 1/2.1.3 of MSC.158(78) for oil tankers

PMA requirement	Means of access requirement	Plated stringers shall, as far as possible, be in alignment with horizontal girders of transverse bulkheads
	IACS interpretation	No interpretation provided
PMA+ requirement		No additional **PMA+** requirements
Additional guidance		• Detailed walkway design guidance is available in Chap. 3, section "Walkways and Ramps"

Fig. 2.6 Access where the vertical distance between the horizontal uppermost stringer and deck hold is ≥6 m (19.5 ft)

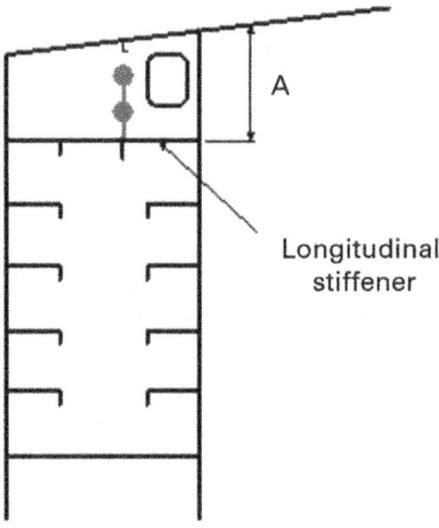

Means of Access Requirements for Table 1/2.2 of MSC.158(78) for Oil Tankers (Access for Distances ≤6 m (19.5 ft) from the Tank Bottom to the Upper Knuckle Point)

Means of access requirements for Table 1/2.2 of MSC.158(78) for oil tankers			
PMA requirement	Means of access requirement	"For bilge hopper sections of which the vertical distance from the tank bottom to the upper knuckle point is 6 m and over, one longitudinal permanent means of access shall be provided for the full length of the tank. It shall be accessible by vertical permanent means of access at both ends of the tank"	
		Dimension		MSC.158(78) requirement
		A	Distance from the top of the bilge hopper	≥1600 mm (63.0 in) and ≤3 m (10.0 ft)
		B	Hopper height	≥6 m (19.5 ft)

(continued)

2.4 Specific Means of Access Requirements for Oil Tankers

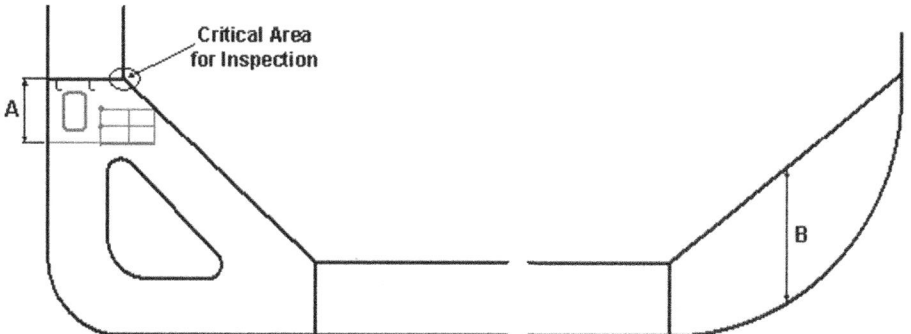

Fig. 2.7 Access for bilge hopper sections where the vertical distance from the tank bottom to the upper knuckle point is ≥6 m (19.5 ft)

(continued)

Means of access requirements for Table 1/2.2 of MSC.158(78) for oil tankers			
	IACS interpretation	"Permanent means of access between the longitudinal continuous permanent means of access and the bottom of the space is to be provided The height of a bilge hopper tank located outside of the parallel part of vessel is to be taken as the maximum of the clear vertical distance measured from the bottom plating to the hopper plating of the tank The foremost and aft most bilge hopper ballast tanks with raised bottom, of which the height is 6 m (19.5 ft) and over, a combination of transverse and vertical MA for access to the upper knuckle point for each transverse web is to be accepted in place of the longitudinal permanent means of access"	
PMA+ requirement		Dimension	**PMA+** requirements
		A	Distance from the top of the bilge hopper	≥2130 mm (84.0 in) and ≤3 m (10.0 ft)
Additional guidance		• Figure 2.7, "Access for bilge hopper sections where the vertical distance from the tank bottom to the upper knuckle point is ≥6 m" is provided to help illustrate this requirement • Distance "A" from the top of the bilge hopper is measured from the underside of the plating to the top of the platform • Detailed walkway design guidance is available in Chap. 3, section "Walkways and Ramps" • Detailed hatch design guidance is available in Chap. 5, "Hatches"	

Means of Access Requirements for Table 1/2.2.1 of MSC.158(78) for Oil Tankers (Continuous Longitudinal Means of Access)

Means of access requirements for Table 1/2.1.2 of MSC.158(78) for oil tankers			
PMA requirement	Means of access requirement	"The longitudinal continuous permanent means of access may be installed at a minimum 1.6 m to maximum 3 m from the top of the bilge hopper section. In this case, a platform extending the longitudinal continuous permanent means of access in way of the web frame may be used to access the identified structural critical areas"	
		Dimension	MSC.158(78) requirement
		A Distance from the top of the bilge hopper	≥1600 mm (63.0 in) and ≤3 m (10.0 ft)
		B Hopper height	≥6 m (19.5 ft)
	IACS interpretation	"The bilge hopper tanks at fore and aft of cargo area narrow due to raised bottom plating and the actual vertical distance from the bottom of the tank to hopper plating of the tank is more appropriate to judge if portable means of access could be utilised for the purpose"	
PMA+ requirement		Dimension	**PMA+** requirements
		A Distance from the top of the bilge hopper	≥2130 mm (84.0 in) and ≤3 m (10.0 ft)
Additional guidance	• Figure 2.8, "Access for bilge hopper tank" is provided to help illustrate this requirement • Distance "A" from the top of the bilge hopper is measured from the underside of the plating to the top of the platform • In this figure, "B" refers to the height of bilge hopper tank. This is the maximum of the clear vertical distance from bottom plating to hopper plating • Detailed walkway design guidance is available in Chap. 3, section "Walkways and Ramps" • Detailed hatch design guidance is available in Chap. 5, "Hatches"		

Means of Access Requirements for Table 1/2.2.2 of MSC.158(78) for Oil Tankers (Continuous Longitudinal Means of Access)

2.4 Specific Means of Access Requirements for Oil Tankers

Fig. 2.8 Access for bilge hopper tank

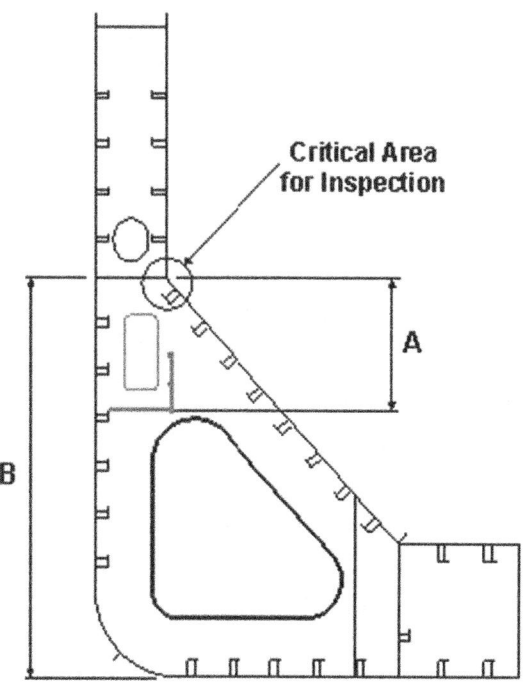

Means of access requirements for Table 1/2.2.2 of MSC.158(78) for oil tankers

PMA requirement	Means of access requirement	"Alternatively, the continuous longitudinal permanent means of access may be installed at a minimum of 1.2 m (4.0 ft) below the top of the clear opening of the web ring allowing a use of portable means of access to reach identified structural critical areas"	
		Dimension	MSC.158(78) requirement
		A Distance below web ring	≥1200 mm (47.0 in)
	IACS interpretation	"In the foremost or aft most bilge hopper tanks where the vertical distance is 6 m or over but installation of longitudinal permanent means of access is not practicable permanent means of access of combination of transverse and vertical ladders provides an alternative means of access to the upper knuckle point"	
PMA+ requirement		Dimensions	**PMA+** requirements
		A Distance below web ring	≥1500 mm (59.0 in)

(continued)

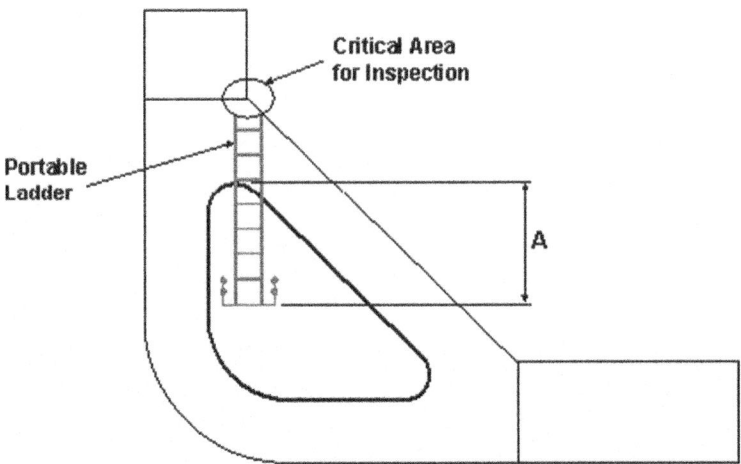

Fig. 2.9 Alternative means of access

(continued)

Means of access requirements for Table 1/2.2.2 of MSC.158(78) for oil tankers	
Additional guidance	• Figure 2.9, "Alternative means of access" is provided to help illustrate this requirement • Detailed guidance on the design and selection of alternative means of access is available in Chap. 6, "Alternative Means of Access" • Detailed walkway design guidance is available in Chap. 3, section "Walkways and Ramps"

Means of Access Requirements for Table 1/2.3 of MSC.158(78) for Oil Tankers (Access for Distances <6 m (19.5 ft) from the Tank Bottom to the Upper Knuckle Point)

Means of access requirements for Table 1/2.3 of MSC.158(78) for oil tankers		
PMA requirement	Means of access requirement	"Where the vertical distance referred to in 2.2 is less than 6 m, alternative means as defined in paragraph 3.9 of the Technical provisions or portable means of access may be utilised in lieu of the permanent means of access. To facilitate the operation of the alternative means of access, in-line openings in horizontal stringers shall be provided. The openings shall be of an adequate diameter and shall have suitable protective railings"

(continued)

(continued)

Means of access requirements for Table 1/2.3 of MSC.158(78) for oil tankers

		Dimension		MSC.158(78) requirement
		A	Handrail height	$\geq$1000 mm (39.5 in)
		B	Distance between longitudinal stringers	<6 m (19.5 ft)
	IACS interpretation	No interpretation provided		
PMA+ requirement		No additional **PMA+** requirements		
Additional guidance		<ul><li>Figure 2.10, "Access for vertical distances <6 m (19.5 ft) from the tank bottom to the upper knuckle point" is provided to help illustrate these requirements</li><li>Detailed walkway design guidance is available in Chap. 3, section "Walkways and Ramps"</li><li>Detailed hatch design guidance is available in Chap. 5, "Hatches"</li><li>Detailed ladder design guidance is available in Chap. 4, "Vertical Ladders, Inclined Ladders and Handles"</li><li>Detailed guidance on the design and selection of alternative means of access is available in Chap. 6, "Alternative Means of Access"</li><li>With the Administration's approval, "reasonable deviations" may be applied to facilitate this means of access. IACS UI (SC) 191 has interpreted this to be no more than 10% for vertical distances exceeding 6 m (19.5 ft)</li></ul>		

2.5 Specific Means of Access Requirements for Bulk Carriers

Means of Access Requirement for Table 2/1.1 of MSC.158(78) for Bulk Carriers (Cargo Holds—Access to Underdeck Structure)

Fig. 2.10 Access for vertical distances <6 m (19.5 ft) from the tank bottom to the upper knuckle point

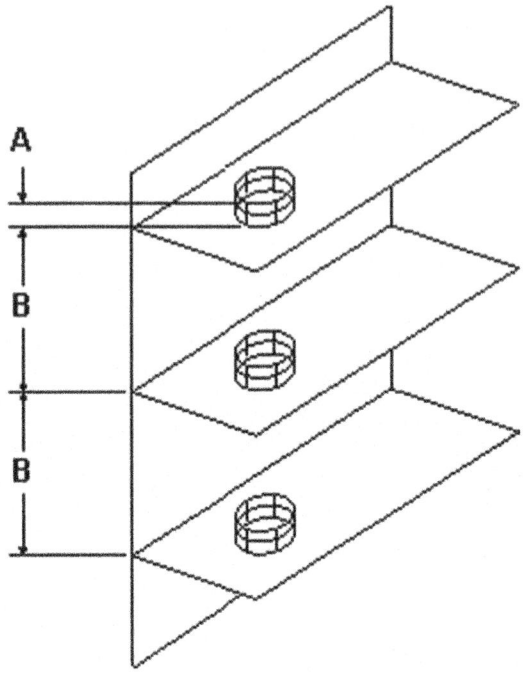

Means of access requirements for Table 2/1.1 of MSC.158(78) for bulk carriers

PMA requirement	Means of access requirement	"Permanent means of access shall be fitted to provide access to the overhead structure at both sides of the cross deck and in the vicinity of the centreline. Each means of access shall be accessible from the cargo hold access or directly from the main deck and installed at a minimum of 1.6 m to a maximum of 3 m below the deck"	
		Dimension	MSC.158(78) requirement
		A Distance below deckhead	≥1600 mm (63.0 in) and ≤3 m (10.0 ft)
	IACS interpretation		
PMA+ requirement		Dimension	**PMA+** requirement
		A Distance below deckhead	≥2130 mm (84.0 in) and ≤3 m (10.0 ft)

(continued)

2.5 Specific Means of Access Requirements for Bulk Carriers

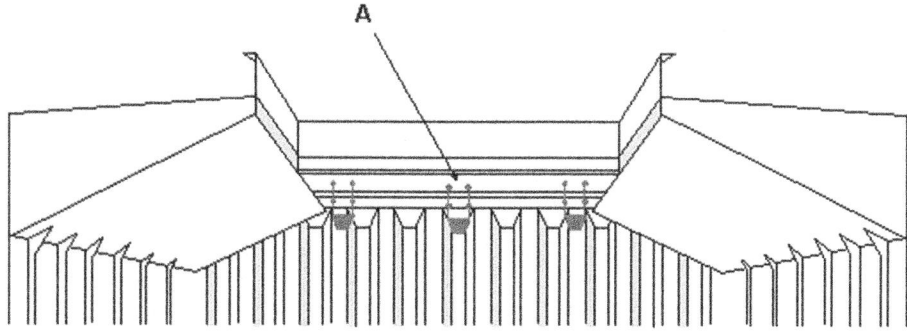

Fig. 2.11 Access to underdeck structures ≥17 m (56.0 ft) in height

(continued)

Means of access requirements for Table 2/1.1 of MSC.158(78) for bulk carriers	
Additional guidance	• Figure 2.11, "Access to underdeck structures ≥17 m (56.0 ft) in height" is provided to help illustrate this requirement • Distance "A" below the deckhead is measured from the underside of the deck plating to the top of the platform • Detailed walkway design guidance is available in Chap. 3, section "Walkways and Ramps"

Means of Access Requirement for Table 2/1.2 of MSC.158(78) for Bulk Carriers (Cargo Holds—Access to Underdeck Structures)

Means of access requirements for Table 2/1.2 of MSC.158(78) for bulk carriers				
PMA requirement	Means of access requirement	"An athwartship permanent means of access fitted on the transverse bulkhead at a minimum 1.6 m to a maximum 3 m below the cross-deck head is accepted as equivalent to 1.1"		
		Dimension		MSC.158(78) requirement
		A	Distance below deckhead	≥1600 mm (63.0 in) and ≤3 m (10.0 ft)
	IACS interpretation	No interpretation provided		
PMA+ requirement		Dimensions		**PMA+** requirement
		A	Distance below deckhead	

(continued)

(continued)

Means of access requirements for Table 2/1.2 of MSC.158(78) for bulk carriers	
Additional guidance	• Figure 2.12, "Athwartship access fitted on the transverse bulkhead ≥ 17 m (56.0 ft) in height" is provided to help illustrate this requirement • Distance "A" below the deckhead is measured from the underside of the deck plating to the top of the platform • Detailed walkway design guidance is available in Chap. 3, section "Walkways and Ramps"

Means of Access Requirement for Table 2/1.3 of MSC.158(78) for Bulk Carriers (Cargo Holds—Access to Underdeck Structure)

Means of access requirements for Table 2/1.3 of MSC.158(78) for bulk carriers		
PMA requirement	Means of access requirement	"Access to the permanent means of access to overhead structure of the cross deck may also be via the upper stool"
	IACS interpretation	"Particular attention is to be paid to preserve the structural strength in way of access opening provided in the main deck or cross deck"
PMA+ requirement		No additional **PMA+** requirements
Additional guidance		• Detailed walkway design guidance is available in Chap. 3, section "Walkways and Ramps"

Means of Access Requirement for Table 2/1.4 of MSC.158(78) for Bulk Carriers (Cargo Holds—Access to Underdeck Structure)

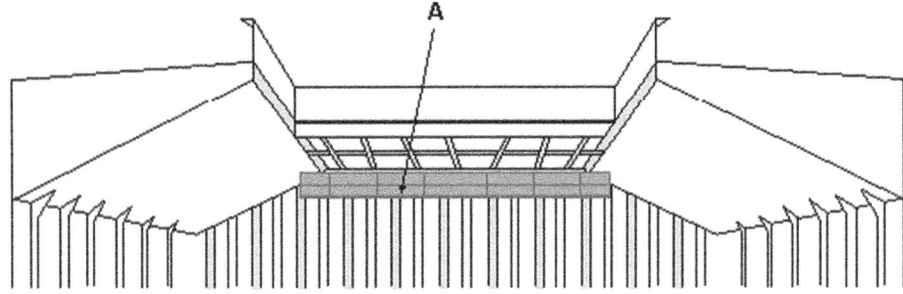

Fig. 2.12 Athwartship access fitted on the transverse bulkhead ≥ 17 m (56.0 ft) in height

2.5 Specific Means of Access Requirements for Bulk Carriers

Means of access requirements for Table 2/1.4 of MSC.158(78) for bulk carriers

PMA requirement	Means of access requirement	"Ships having transverse bulkheads with full upper stools with access from the main deck which allows monitoring of all framing and plates from inside, do not require permanent means of access of the cross deck"
	IACS interpretation	"Full upper stools' are understood to be stools with a full extension between top side tanks and between hatch end beams"
PMA+ requirement		No **PMA+** requirement
Additional guidance		• Detailed walkway design guidance is available in Chap. 3, section "Walkways and Ramps"

Means of Access Requirement for Table 2/1.5 of MSC.158(78) for Bulk Carriers (Cargo Holds—Access to Underdeck Structure)

Means of access requirements for Table 2/1.5 of MSC.158(78) for bulk carriers

PMA requirement	Means of access requirement	"Alternatively, movable means of access may be utilised for access to the overhead structure of cross deck if its vertical distance is 17 m or less above the tank top"
	IACS interpretation	"The movable means of access to the underdeck structure of cross deck need not necessarily be carried on board the vessel. It is sufficient if it is made available when needed. The requirement for bulk carrier cross deck structure is also considered applicable to ore carriers"
PMA+ requirement		No **PMA+** requirement
Additional guidance		• This movable means of access shall not be a vertical ladder, except for heights under 6 m (19.5 ft) • Detailed guidance on the design and selection of alternative means of access is available in Chap. 6, "Alternative Means of Access" • Detailed ladder design guidance is available in Chap. 4, "Vertical Ladders, Inclined Ladders and Handles"

Means of Access Requirement for Table 2/1.6 of MSC.158(78) for Bulk Carriers (Cargo Holds—Access to Vertical Structures)

Means of access requirements for Table 1/2.1.6 of MSC.158(78) for bulk carriers

PMA requirement	Means of access requirement	"Permanent means of vertical access shall be provided in all cargo holds and built into the structure to allow for an inspection of a minimum of 25% of the total number of hold frames port and starboard equally distributed throughout the hold including at each end in way of transverse bulkheads. But in no circumstance shall this arrangement be less than 3 permanent means of vertical access fitted to each side (fore and aft ends of hold and mid-span). Permanent means of vertical access fitted between two adjacent hold frames is counted for an access for the inspection of both hold frames. A means of portable access may be used to gain access over the sloping plating of lower hopper ballast tanks"	
		Dimension	MSC.158(78) requirement
		A Distance between rungs	$\leq$6 m (19.5 ft)
	IACS interpretation	• "The maximum vertical distance of the rungs of vertical ladders for access to hold frames is to be 350 mm • If safety harness is to be used, means should be provided for connecting the safety harness in suitable places in a practical way"	
PMA+ requirement		Dimension	**PMA+** requirement
		A Distance between rungs	$\geq$275 mm (11.0 in) and $\leq$300 mm (12.0 in)
Additional guidance		• Figure 2.13, "Bulk carriers having transverse bulkheads with full upper stools $\geq$17 m (56.0 ft) in height" is provided to help illustrate this requirement • Permanent access can be a vertical ladder or horizontal bars/rungs in the hold frames • Ladder rung-to-rung distance consistent for the full run of the ladder • Detailed ladder design guidance is available in Chap. 4, "Vertical Ladders, Inclined Ladders and Handles"	

Means of Access Requirement for Table 2/1.7 of MSC.158(78) for Bulk Carriers (Cargo Holds—Access to Vertical Structures)

2.5 Specific Means of Access Requirements for Bulk Carriers

Fig. 2.13 Bulk carriers having transverse bulkheads with full upper stools ≥17 m (56.0 ft) in height

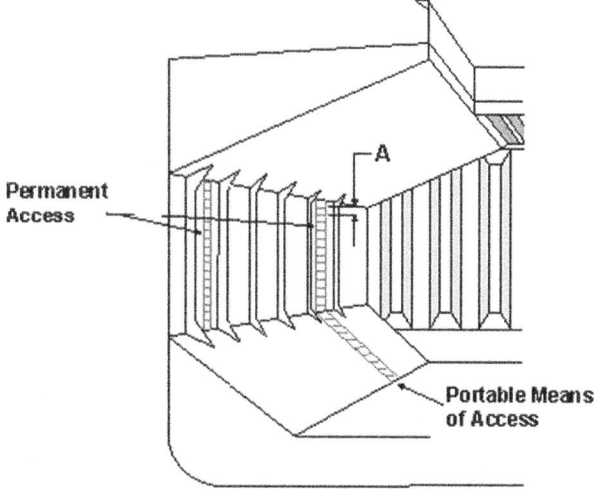

Means of access requirements for Table 2/1.7 of MSC.158(78) for bulk carriers

PMA requirement	Means of access requirement	"In addition to 1.6, portable or movable means of access shall be utilised for access to the remaining hold frames up to their upper brackets and transverse bulkheads"
	IACS interpretation	"Portable, movable or alternative means of access also is to be applied to corrugated bulkheads"
PMA+ requirement		No **PMA+** requirement
Additional guidance		• Detailed guidance on the design and selection of alternative means of access is available in Chap. 6, "Alternative Means of Access"

Means of Access Requirement for Table 2/1.8 of MSC.158(78) for Bulk Carriers (Cargo Holds—Access to Vertical Structures)

Means of access requirements for Table 2/1.8 of MSC.158(78) for bulk carriers

PMA requirement	Means of access requirement	"Portable or movable means of access may be utilised for access to hold frames up to their upper bracket in place of the permanent means required in 1.6. These means of access shall be carried on board the ship and readily available for use"
	IACS interpretation	Readily available means: "Able to be transported to location in cargo hold and safely erected by ship's staff"

(continued)

(continued)

Means of access requirements for Table 2/1.8 of MSC.158(78) for bulk carriers	
PMA+ requirement	No **PMA+** requirement
Additional guidance	Detailed guidance on the design and selection of alternative means of access is available in Chap. 6, "Alternative Means of Access"

Means of Access Requirement for Table 2/1.9 of MSC.158(78) for Bulk Carriers (Cargo Holds—Access to Vertical Structures)

Means of access requirements for Table 2/1.9 of MSC.158(78) for bulk carriers		
PMA requirement	Means of access requirement	"The width of vertical ladders for access to hold frames shall be at least 300 mm, measured between stringers"
	IACS interpretation	No interpretation provided
PMA+ requirement		No **PMA+** requirement
Additional guidance		Detailed ladder design guidance is available in Chap. 4, "Vertical Ladders, Inclined Ladders and Handles"

Means of Access Requirement for Table 2/1.10 of MSC.158(78) for Bulk Carriers (Cargo Holds—Access to Vertical Structures)

Means of access requirements for Table 2/1.10 of MSC.158(78) for bulk carriers		
PMA requirement	Means of access requirement	"A single vertical ladder over 6 m in length is acceptable for the inspection of the hold side frames in a single skin construction"
	IACS interpretation	No interpretation provided
PMA+ requirement		No **PMA+** requirement
Additional guidance		• Climber safety devices (rails) shall be used for vertical ladders over 6 m (19.5 ft) • Detailed ladder design guidance is available in Chap. 4, "Vertical Ladders, Inclined Ladders and Handles"

Means of Access Requirement for Table 2/1.11 of MSC.158(78) for Bulk Carriers (Cargo Holds—Access to Vertical Structures)

Means of access requirements for Table 2/1.11 of MSC.158(78) for bulk carriers

PMA requirement	Means of access requirement	"For double-side skin construction no vertical ladders for the inspection of the cargo hold surfaces are required. Inspection of this structure should be provided from within the double hull space"
	IACS interpretation	No interpretation provided
PMA+ requirement		No **PMA+** requirement
Additional guidance		• Detailed walkway design guidance is available in Chap. 3, section "Walkways and Ramps"

Means of Access Requirement for Table 2/2.1 of MSC.158(78) for Bulk Carriers (Ballast Tanks—Top Side Tanks)

Means of access requirements for Table 2/2.1 of MSC.158(78) for bulk carriers

PMA requirement	Means of access requirement	"For each topside tank of which the height is 6 m and over, one longitudinal continuous permanent means of access shall be provided along the side shell webs and installed at a minimum of 1.6 m to a maximum of 3 m below deck with a vertical access ladder in the vicinity of each access to that tank"		
		Dimension		MSC.158(78) requirement
		A	Distance below deckhead	≥1.6 m (63 in) and ≤3 m (10.0 ft)
	IACS interpretation			
PMA+ requirement		Dimension		**PMA+** requirement
		A	Distance below deckhead	≥2130 mm (84.0 in) and ≤3 m (10.0 ft)

(continued)

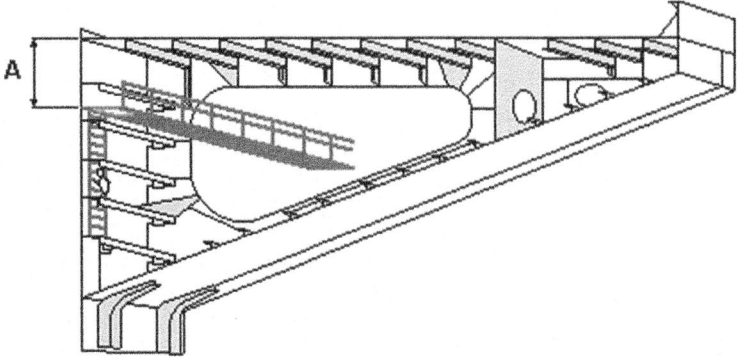

Fig. 2.14 Upper topside tank access with a height >6 m (19.5 ft)

(continued)

Means of access requirements for Table 2/2.1 of MSC.158(78) for bulk carriers	
Additional guidance	• Figure 2.14, "Upper topside tank access with a height >6 m" is provided to help illustrate this requirement • Distance "A" below the deckhead is measured from the underside of the plating to the top of the platform • Detailed walkway design guidance is available in Chap. 3, section "Walkways and Ramps" • Detailed ladder design guidance is available in Chap. 4, "Vertical Ladders, Inclined Ladders and Handles"

Means of Access Requirement for Table 2/2.2 of MSC.158(78) for Bulk Carriers (Ballast Tanks—Top Side Tanks)

Means of access requirements for Table 2/2.2 of MSC.158(78) for bulk carriers				
PMA requirement	Means of access requirement	"If no access holes are provided through the transverse webs within 600 mm (23.5 in) of the tank base and the web frame rings have a web height greater than 1 m (39.5 in) in way of side shell and sloping plating, then step rungs/grab rails shall be provided to allow safe access over each transverse web frame ring"		
		Dimension		MSC.158(78) requirement
		A	Step rungs/grab rails	Not provided
	IACS interpretation	No interpretation provided		
PMA+ requirement		Dimension		**PMA+** requirement

(continued)

2.5 Specific Means of Access Requirements for Bulk Carriers

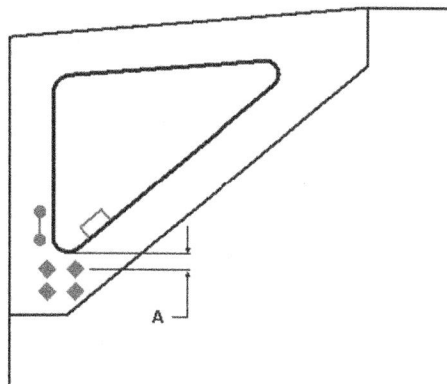

Fig. 2.15 Transverse web access for bulk carriers

(continued)

Means of access requirements for Table 2/2.2 of MSC.158(78) for bulk carriers

	A	Step rungs/grab rails	≥275 mm (11.0 in) and ≤300 mm (12.0 in)
Additional guidance	• Figure 2.15, "Transverse web access for bulk carriers" is provided to help illustrate this requirement • Detailed guidance on the design of handles is available in Chap. 4, section "Handles" • Detailed guidance for individual stairs is available in Chap. 4, section "Individual Rung Ladders"		

Means of Access Requirement for Table 2/2.3 of MSC.158(78) for Bulk Carriers (Ballast Tanks—Top Side Tanks)

Means of access requirements for Table 2/2.3 of MSC.158(78) for bulk carriers

PMA requirement	Means of access requirement	"Three permanent means of access, fitted at the end bay and middle bay of each tank, shall be provided spanning from tank base up to the intersection of the sloping plate with the hatch side girder. The existing longitudinal structure may be used as part of this means of access"
	IACS interpretation	"If the longitudinal structures on the sloping plate are fitted outside of the tank a means of access is to be provided"
PMA+ requirement		No additional **PMA+** requirements
Additional guidance		• Detailed guidance on the design of handles is available in Chap. 4, section "Handles"

Means of Access Requirement for Table 2/2.4 of MSC.158(78) for Bulk Carriers (Ballast Tanks—Top Side Tanks)

Means of access requirements for Table 2/2.4 of MSC.158(78) for bulk carriers

PMA requirement	Means of access requirement	"For topside tanks of which the height is less than 6 m, alternative means as defined in paragraph 3.9 of the Technical provisions or portable means may be utilised in lieu of the permanent means of access"
	IACS interpretation	No interpretation provided
PMA+ requirement		No additional **PMA+** requirements
Additional guidance		Detailed guidance on the design and selection of alternative means of access is available in Chap. 6, "Alternative Means of Access"

2.5 Specific Means of Access Requirements for Bulk Carriers

Means of Access Requirement for Table 2/2.5 of MSC.158(78) for Bulk Carriers (Ballast Tanks—Bilge Hopper Tanks)

Means of access requirements for Table 1/2.5 of MSC.158(78) for bulk carriers

PMA requirement	Means of access requirement	"For each bilge hopper tank of which the height is 6 m and over, one longitudinal continuous permanent means of access shall be provided along the side shell webs and installed at a minimum of 1.2 m (47.0 in) below the top of the clear opening of the web ring with a vertical access ladder in the vicinity of each access to the tank"		
		Dimension		MSC.158(78) requirement
		A	Distance below web ring	≥1200 mm (47.0 in)
	IACS interpretation			
PMA+ requirement		Dimension		**PMA+** requirement
		A	Distance below web ring	≥1500 mm (59.0 in)
Additional guidance		• Figure 2.16, "Bilge hopper tank access with a height ≥6 m" is provided to help illustrate these requirements • Detailed walkway design guidance is available in Chap. 3, section "Walkways and Ramps"		

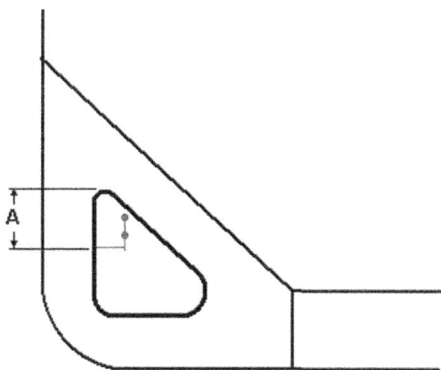

Fig. 2.16 Bilge hopper tank access with a height ≥6 m (19.5 ft)

Means of Access Requirement for Table 2/2.5.1 of MSC.158(78) for Bulk Carriers (Ballast Tanks—Bilge Hopper Tanks)

Means of access requirements for Table 2/2.5.1 of MSC.158(78) for bulk carriers		
PMA requirement	Means of access requirement	"An access ladder between the longitudinal continuous permanent means of access and the bottom of the space shall be provided at each end of the tank"
	IACS interpretation	No interpretation provided
PMA+ requirement		No additional **PMA+** requirements
Additional guidance		• Figure 2.17, "Access ladder between the longitudinal continuous means of access and the bottom of the space" is provided to help illustrate this requirement • Detailed ladder design guidance is available in Chap. 4, "Vertical Ladders, Inclined Ladders and Handles" • Detailed walkway design guidance is available in Chap. 3, section "Walkways and Ramps"

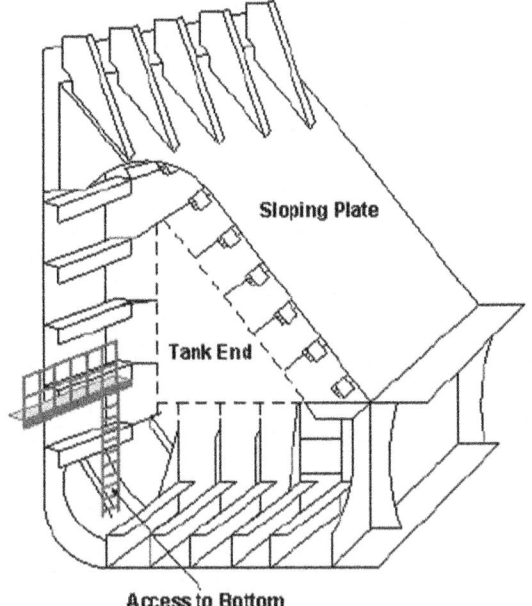

Fig. 2.17 Access ladder between the longitudinal continuous means of access and the bottom of the space

Means of Access Requirement for Table 2/2.5.2 of MSC.158(78) for Bulk Carriers (Ballast Tanks—Bilge Hopper Tanks)

Means of access requirements for Table 2/2.5.2 of MSC.158(78) for bulk carriers

PMA requirement	Means of access requirement	"Alternatively, the longitudinal continuous permanent means of access can be located through the upper web plating above the clear opening of the web ring, at a minimum of 1.6 m below the deck head, when this arrangement facilitates more suitable inspection of identified structurally critical areas. An enlarged longitudinal frame can be used for the purpose of the walkway"		
		Dimension		MSC.158(78) requirement
		A	Distance below deck head	≥1600 mm (63.0 in)
		B	Access width	≥600 mm (23.5 in)
	IACS interpretation	"A wide longitudinal frame of at least 600 mm (23.5 in) clear width may be used for the purpose of the longitudinal continuous permanent means of access." The foremost and aftermost bilge hopper ballast tanks with raised bottom, of which the height is 6 m and over, a combination of transverse and vertical MA for access to the sloping plate of hopper tank connection with side shell plating for each transverse web can be accepted in place of the longitudinal permanent means of access"		
PMA+ requirement		Dimension		PMA+ requirement
		A	Distance below deck head	≥2130 mm (84.0 in)
		B	Access width	700 mm (27.5 in)
Additional guidance		<ul><li>Figure 2.18, "Alternate bilge hopper tank access with a height ≥6 m" is provided to help illustrate these requirements</li><li>Vertical walkway clearance from a **PMA+** perspective shall be measured from the top of the walking platform/surface to the lowest structure directly above the walkway (e.g., stiffener)</li><li>Detailed walkway design guidance is available in Chap. 3, section "Walkways and Ramps"</li></ul>		

Fig. 2.18 Alternate bilge hopper tank access with a height ≥6 m (19.5 ft)

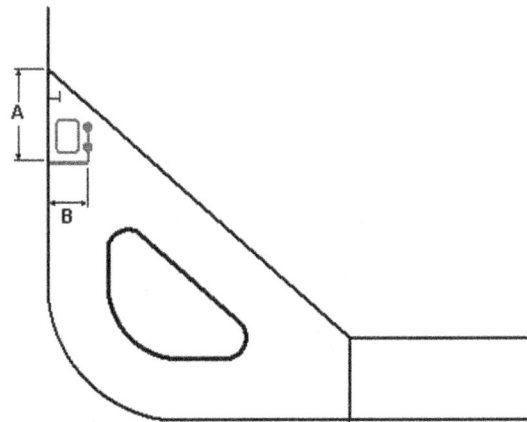

Means of Access Requirement for Table 2/2.5.3 of MSC.158(78) for Bulk Carriers (Ballast Tanks—Bilge Hopper Tanks)

Means of access requirements for Table 2/2.5.3 of MSC.158(78) for bulk carriers				
PMA requirement	Means of access requirement	"For double-side skin bulk carriers, the longitudinal continuous permanent means of access may be installed within 6 m from the knuckle point of the bilge, if used in combination with alternative methods to gain access to the knuckle point"		
		Dimension		MSC.158(78) requirement
		A	Distance below web ring	≥1200 mm (47.0 in)
	IACS interpretation	No interpretation provided		
PMA+ requirement		Dimension		**PMA+** requirement
		A	Distance below web ring	≥1500 mm (49.0 in)
Additional guidance		• Figure 2.19, "Access for foremost and aftmost bilge hopper tanks" are provided to help illustrate these requirements • Vertical walkway clearance from a **PMA+** perspective shall be measured from the top of the walking platform/surface to the lowest structure directly above the walkway (e.g., stiffener) • Detailed walkway design guidance is available in Chap. 3, section "Walkways and Ramps" • Detailed ladder design guidance is available in Chap. 4, "Vertical Ladders, Inclined Ladders and Handles"		

2.5 Specific Means of Access Requirements for Bulk Carriers

Fig. 2.19 Access for foremost and aftmost bilge hopper tanks

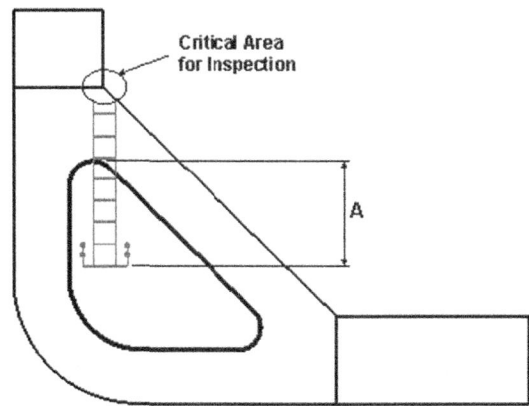

Means of Access Requirement for Table 2/2.6 of MSC.158(78) for Bulk Carriers (Ballast Tanks—Bilge Hopper Tanks)

Means of access requirements for Table 2/2.6 of MSC.158(78) for bulk carriers		
PMA requirement	Means of access requirement	"If no access holes are provided through the transverse ring webs within 600 mm (23.5 in) of the tank base and the web frame rings have a web height greater than 1 m (39.5 in) in way of side shell and sloping plating, then step rungs/grab rails shall be provided to allow safe access over each transverse web frame ring"
	IACS interpretation	"The height of web frame rings should be measured in way of side shell and tank base"
PMA+ requirement		No additional **PMA+** requirements
Additional guidance		• Detailed ladder design guidance is available in Chap. 4, "Vertical Ladders, Inclined Ladders and Handles" • Detailed hatch design guidance is available in Chap. 5, "Hatches"

Means of Access Requirement for Table 2/2.7 of MSC.158(78) for Bulk Carriers (Ballast Tanks—Bilge Hopper Tanks)

Means of access requirements for Table 2/2.7 of MSC.158(78) for bulk carriers

PMA requirement	Means of access requirement	"For bilge hopper tanks of which the height is less than 6 m, alternative means as defined in paragraph 3.9 of the Technical provisions or portable means may be utilised in lieu of the permanent means of access. Such means of access shall be demonstrated that they can be deployed and made readily available in the areas where needed"
	IACS interpretation	No interpretation provided
PMA+ requirement		No additional **PMA+** requirements
Additional guidance		• Detailed guidance on the design and selection of alternative means of access is available in Chap. 6, "Alternative Means of Access"

Means of Access Requirement for Table 2/2.8 of MSC.158(78) for Bulk Carriers (Ballast Tanks—Bilge Hopper Tanks—Double-Skin Side Tanks)

Means of access requirements for Table 2/2.8 of MSC.158(78) for bulk carriers

PMA requirement	Means of access requirement	"Permanent means of access shall be provided in accordance with the applicable sections of Resolution MSC158(78), Table 1"
	IACS interpretation	No interpretation provided
PMA+ requirement		No additional **PMA+** requirements
Additional guidance		• Detailed walkway design guidance is available in Chap. 3, section "Walkways and Ramps" • Detailed ladder design guidance is available in Chap. 4, "Vertical Ladders, Inclined Ladders and Handles" • Detailed hatch design guidance is available in Chap. 5, "Hatches"

Means of Access Requirement for Table 2/2.9 of MSC.158(78) for Bulk Carriers (Ballast Tanks—Fore Peak Tanks)

Means of access requirements for Table 2/2.9 of MSC.158(78) for bulk carriers		
PMA requirement	Means of access requirement	"For fore peak tanks with a depth of 6 m or more at the centreline of the collision bulkhead, a suitable means of access shall be provided for access to critical areas such as the underdeck structure, stringers, collision bulkhead and side shell structure"
	IACS interpretation	No interpretation provided
PMA+ requirement		No additional **PMA+** requirements
Additional guidance		• Detailed walkway design guidance is available in Chap. 3, section "Walkways and Ramps" • Detailed ladder design guidance is available in Chap. 4, "Vertical Ladders, Inclined Ladders and Handles" • Detailed hatch design guidance is available in Chap. 5, "Hatches"

Means of Access Requirement for Table 2/2.9.1 of MSC.158(78) for Bulk Carriers (Ballast Tanks—Fore Peak Tanks)

Means of access requirements for Table 2/2.9.1 of MSC.158(78) for bulk carriers		
PMA requirement	Means of access requirement	"Stringers of less than 6 m in vertical distance from the deck head or a stringer immediately above are considered to provide suitable access in combination with portable means of access"
	IACS interpretation	No interpretation provided
PMA+ requirement		No additional **PMA+** requirements
Additional guidance		• Detailed ladder design guidance is available in Chap. 4, "Vertical Ladders, Inclined Ladders and Handles" • Detailed walkway design guidance is available in Chap. 3, section "Walkways and Ramps" • Detailed hatch design guidance is available in Chap. 5, "Hatches"

Means of Access Requirement for Table 2/2.9.2 of MSC.158(78) for Bulk Carriers (Ballast Tanks—Fore Peak Tanks)

Means of access requirements for Table 2/2.9.2 of MSC.158(78) for bulk carriers		
PMA requirement	Means of access requirement	"In case the vertical distance between the deck head and stringers, stringers or the lowest stringer and the tank bottom is 6 m or more, alternative means of access as defined in paragraph 3.9 of the Technical provisions shall be provided"
	IACS interpretation	No interpretation provided
PMA+ requirement		No **PMA+** requirement requirements
Additional guidance		• Detailed guidance on the design and selection of alternative means of access is available in Chap. 6, "Alternative means of access" • Detailed walkway design guidance is available in Chap. 3, section "Walkways and Ramps" • Detailed ladder design guidance is available in Chap. 4, "Vertical Ladders, Inclined Ladders and Handles" • Detailed hatch design guidance is available in Chap. 5, "Hatches"

3 Walkways, Ramps and Work Platforms

3.1 General

This Section contains design guidance for walkways, ramps and work platforms. The guidance included in the figures and tables below provides the design attribute (application) and the IMO (**PMA**) and **PMA+** dimension requirements. There are instances where IMO means of access requirements do not provide specific design dimensioning. In these instances, the **PMA+** dimensions may be used as guidance.

Design Loads

IMO requirements state that the construction and materials of all means of access and their attachment to the vessel's structure shall be to the satisfaction of the Flag State Administration and that the means of access shall be of "substantial construction" and "adequate strength and stiffness". The IACS UI SC191 definition of substantial construction is as follows: "Substantial construction is taken to refer to the as-designed strength as well as the residual strength during the service life of the vessel. Durability of passageways together with guardrails shall be verified by the initial corrosion protection and inspection and maintenance during services."

The design loads listed in the following Subparagraphs are provided as additional quantitative requirements for the **PMA+** notation. Where requirements for design loads, specified by other regulatory bodies (e.g., Flag State Administrations and Port State authorities), are greater than these design loads, those requirements take precedence over this text. This text defines "design load" as the maximum intended load, being the total of all loads including the weight of the personnel, materials and equipment, including the means of access structure.

Guardrails

Guardrails should be able to withstand anticipated loads that are not less than 90 kg (200 lbs) at any point and in any direction when applied to the top rail.

Walkways and Work Platforms

The minimum design loads for the landings, walkways and working platforms are:

- 2.0 kN/m^2 (0.29 lbf/in^2) under uniform load for the structure; and
- 1.5 kN (337 lbf) concentrated load applied in the most unfavourable position over a concentrated load area of 200 mm × 200 mm (8.0 in × 8.0 in) for the flooring.

When loaded with the design load, the deflection of the flooring shall not exceed 1/200th of the span and the difference between the loaded and adjacent unloaded flooring shall not exceed 4 mm (0.16 in) in height.

3.2 Walkways and Ramps

This section includes general principles as well as the design requirements for the arrangement of walkways and ramps, and the provision of guardrails and handrails.

General Principles

The principles listed below apply to the design of walkways and ramps and are not represented in the following figures or tables (Fig. 3.1):

- Walkway width is dependent on factors such as the demand of any tasks performed on or from the walkway, frequency of use and the number of workers using the walkway at the same time,
- Guardrails shall be provided at the exposed side of any walking or standing surface that is 600 mm (23.5 in) or higher above the adjacent surface and where a person could fall from the upper to the lower surface,
- Ramps are best used with changes in vertical elevations of less than 600 mm (23.5 in) but may be used for any height provided that the angle of inclination to the horizontal complies with Fig. 3.2 "Walkway and ramp design,"
- Toe boards shall be considered on elevated walkways, platforms and ramps,
- Permanent means of access shall as far as possible be integral to the structure of the vessel, thus verifying that they are robust and at the same time contributing to the overall strength of the structure of the vessel,

3.2 Walkways and Ramps

	Dimension	Guideline
A	Height of toe board	100 mm (4.0 in)
B	Gap between toe board and surface	6 mm (0.25 in)

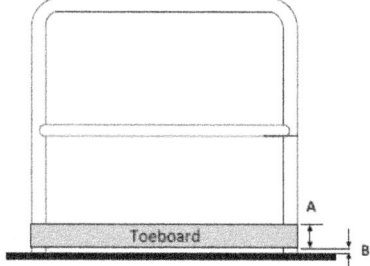

Fig. 3.1 Toe board dimensions

- Where stays are provided for supporting stanchions, they shall be fitted so as not to obstruct safe passage,
- Stanchion scantlings can be formed of flat or round bar. Refer to Fig. 3.2, "Walkway and ramp design,"
- Brackets joining the guardrail stanchions to the means of access shall be oriented in a way to avoid causing a trip hazard (e.g., parallel to direction of the walkway),
- No impediments or tripping hazards shall intrude into the transit space (for example, electrical boxes, valves, actuators, or piping); and
- No impediments or tripping hazards shall impede use of a walkway or ramp (for example, piping runs, hatch covers, deck impediments (e.g., through bolts) or combings/retention barriers).

Toe Boards

The use of toe boards on elevated walkways and platforms is a standard safety and ergonomics practice. Toe boards help prevent a worker's foot, tools, parts and equipment from slipping or falling off the edge of an elevated walkway or platform. However, the use of toe boards on walkways or platforms used for inspection in cargo tanks and holds could inhibit and delay the safe and complete off-load of cargo as well as create potential hazards on the walkways. Toe boards could retain cargo in bulk carriers as well as retain sludge in oil tankers. Each of these instances could present stepping, tripping and slipping hazards to workers who have to clear, clean or work from the walkways or platforms. As a result, **PMA** and **PMA+** requirements do not require the use of toe

	Dimension	MSC.158(78)/UI SC 191 (PMA) *requirement*	PMA+ *requirement*
A	Walkway width	≥ 600 mm (23.5 in)	≥ 710 mm (28.0 in)
B	Distance behind handrail and any obstruction	No specific requirement	≥ 75 mm (3.0 in)
C	Gaps between two handrail sections or other structural members (refer to Chapter 3, Figures 5 and 6)	≤ 50 mm (2.0 in)	No additional requirement
D	Span between to handrail stanchions	≤ 3.0 m (10.0 ft)	≤ 2.4 m (8.0 ft)
E	Diameter of handrail	No specific requirement	≥ 40 mm (1.5 in) and ≤ 50 mm (2.0 in)
F	Height of handrail (measured to the top of the handrail)	≥ 1,000 mm (39.5 in)	1,070 mm (42.0 in)
G	Height of intermediate rail (measured from the bottom of the intermediate rail to the walking surface)	500 mm (19.5 in)	No additional requirement
H	Maximum distance between the adjacent stanchions across handrail gaps (refer to Chapter 3, Figures 5 and 6)	≤ 350 mm (14.0 in)	No additional requirement
I	Distance below the overhead structure (measured from the underside of the deck plating to the top of the platform)	≥ 1,600 mm (63.0 in) And ≤ 3 m (10.0 ft)	≥ 2,130 mm (84.0 in) and ≤ 3 m (10.0 ft)
J	The maximum opening in a walkway grating under which the presence of persons is expected (i.e., persons are working and does not include occasional passage)	No specific requirement	≤ 22 mm (0.9 in)
	The maximum opening in a walkway grating under which the presence of persons is not expected	No specific requirement	≤ 35 mm (1.7 in)
K	Ramp (sloping structure) surface	Non-skid construction	Non-skid construction and have a coefficient of friction (COF) of ≥ 0.6 when wet
θ	Ramp angle of inclination	≥ 5 degrees	≥ 5 degrees and ≤ 15 degrees

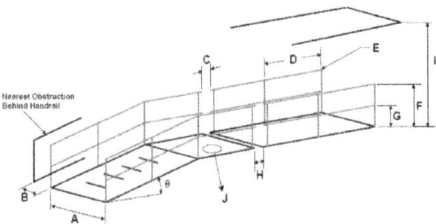

Fig. 3.2 Walkway and ramp design

boards, but their use should be considered. Additional consideration should be given to where toe boards will prove helpful depending on the nature of the cargo and the toe board's potential susceptibility to damage or deterioration from the cargo.

Note: This is a departure from safety and ergonomics practices but is acceptable for means of access for inspection purposes in cargo tanks and holds. If toe boards are to be used, refer to Fig. 3.1, *"Toe board dimensions".*

Walkway and Ramp Design

The dimensions relating to the design of walkways and ramps are presented in Fig. 3.2, "Walkway and ramp design" and Fig. 3.3, "Web frame walkways" (Figs. 3.4 and 3.5).

3.3 Work Platforms

Work platforms shall be provided at locations where personnel must perform tasks that cannot be easily accomplished by reaching from an existing standing surface.

	Dimension	*MSC.158(78)/UI SC 191* **(PMA)** *requirement*	**PMA+** *requirement*
A	Walkway width around a web frame	≥ 450 mm (17.5 in)	≥ 500 mm (19.5 in)

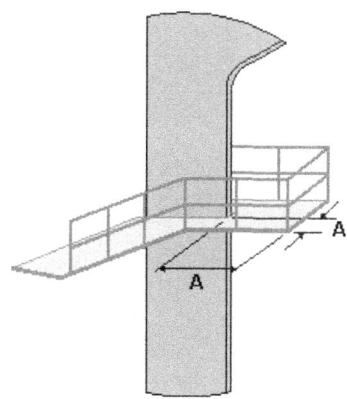

Fig. 3.3 Web frame walkways

	Dimension	MSC.158(78)/UI SC 191 (PMA) *requirement*	PMA+ *requirement*
A	Gaps between two handrail sections	≤ 50 mm (2.0 in)	No additional requirement, but no gaps are preferred
B	Distance between adjacent stanchions	≤ 550 mm (21.5 in)	No additional requirement

+ = *Other handrail measurements apply (e.g., handrail heights and span between handrail stanchions)*

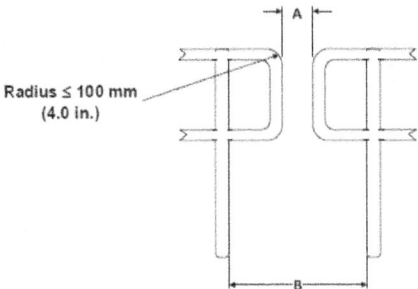

Fig. 3.4 Discontinuous handrail where top and mid rails are connected[+]

General Principles

The principles listed below apply to the design of work platforms:

- Platforms shall be of sufficient size (refer to Fig. 3.6, "Work platform dimensions") to accommodate the task and allow for placement of any required tools, spare parts, or equipment; and
- Work platforms more than 600 mm (23.5 in) above the surrounding surface shall be provided with guard rails and handrails as described in the section "Walkways and Ramps."

3.3 Work Platforms

	Dimension	MSC.158(78)/UI SC 191 (PMA) requirement	PMA+ requirement
A	Gaps between two handrail sections	≤ 50 mm (2.0 in)	No additional requirement, but no gaps are preferred
B	Distance between adjacent stanchions	≤ 550 mm (21.5 in)	No additional requirement

+ = *Other handrail measurements apply (e.g., handrail heights and span between handrail stanchions)*

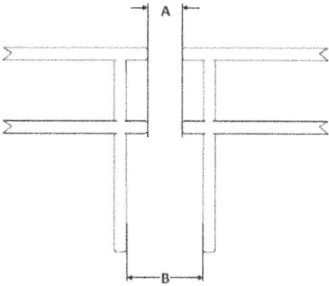

Fig. 3.5 Discontinuous handrails where top and mid rails are not connected[+]

	Dimension	MSC.158(78)/UI SC 191 (PMA) requirement	PMA+ requirement
A	Gaps between two handrail sections	No specific requirement	≥ 750 mm (29.5 in)
	Work platform width (if only used for standing)	No specific requirement	≥ 380 mm (15.0 in)
B	Distance between adjacent stanchions	No specific requirement	≥ 925 mm (36.5 in)
	Work platform length (if only used for standing)	No specific requirement	≥ 450 mm (17.5 in)

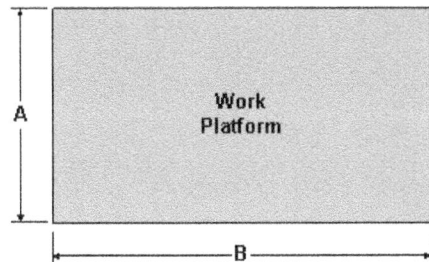

Fig. 3.6 Work platform dimensions

Vertical Ladders, Inclined Ladders and Handles

4.1 General

This chapter contains guidance related to the design of the different attributes of vertical ladders, inclined ladders, individual rung ladders and handles. The guidance included in the figures and tables below provides the design attribute (application) and the IMO (**PMA**) and **PMA+** dimension requirements. There are a few instances where IMO means of access requirements do not provide specific design dimensioning. In these instances, the **PMA+** dimensions may be used as guidance.

Design Loads

IMO requirements state that the construction and materials of all means of access and their attachment to the vessel's structure should be to the satisfaction of the Administration and that the means of access should be of "substantial construction" and "adequate strength and stiffness." The IACS UI SC191 definition of substantial construction is as follows: "Substantial construction is taken to refer to the as-designed strength as well as the residual strength during the service life of the vessel. Durability of passageways together with guardrails should be verified by the initial corrosion protection and inspection and maintenance during services."

The design loads listed in the following paragraphs are provided as additional quantitative requirements for the **PMA+** notation. Where requirements for design loads, specified by other regulatory bodies (e.g., Flag State Administrations and Port State authorities), are greater than these design loads, those requirements take precedence over the guidance provided in this textbook.

This text defines "design load" as the maximum intended load, being the total of all loads including the weight of the personnel, materials, equipment and means of access structure and should be as follows.

Guardrails/Handrails

Guardrails and handrails should withstand anticipated loads but not less than 90 kg (200 lbs) at any point and in any direction when applied to the top rail.

Vertical Ladders

For vertical ladders, the design load should be determined by the anticipated usage of the ladder but should not be less than a single concentrated live load of 90 kg (200 lbs). The weight of the ladder and attached appurtenances together with the design load should be considered in the design of rails and fastenings.

Inclined Ladders

Inclined ladders should be designed and constructed to carry a load of at least three times (five times preferred) the normal load anticipated but never of less strength than to carry safely a moving concentrated load of 225 kg (495 lbs) [455 kg (1000 lbs) preferred].

Landings

The minimum working loads for landings are:

- 2.0 kN/m^2 (0.29 lbf/in^2) under uniform load for the structure,
- 1.5 kN (337 lbf) concentrated load applied in the most unfavourable position over a concentrated load area of 200 mm × 200 mm (8.0 in × 8.0 in) for the flooring; and
- When loaded with the design load, the deflection of the flooring should not exceed 1/200th of the span and the difference between the loaded and adjacent unloaded flooring should not exceed 4 mm (0.16 in) in height.

Use and Selection of Ladders

Stairs, vertical ladders, inclined ladders or ramps should be provided whenever operators or maintainers must change elevation abruptly by more than 300 mm (12.0 in). Guidance relating to ramps can be found in the section "Walkways and Ramps". These structures should also be used, when appropriate, for passage over low objects (e.g., pipes, lines, ridges). Though stairs are the preferred form of access/egress, their practicality in cargo tanks and holds for inspection may be limited. The selection of vertical ladders and inclined ladders should be based on the purpose, frequency of use, and angle of ascent. Refer to Table 4.1, "Selection of access type" for related guidance on angle of inclination.

4.2 Vertical Ladders

Table 4.1 Selection of access type

Dimension	MSC.158(78)/UI SC 191 (PMA) requirement	PMA+requirement
Inclined ladders	<70°	45–60°
Vertical ladders	≥70°	80–90°

4.2 Vertical Ladders

This section contains guidance on the design of vertical ladders and climber safety devices.

General Principles

The considerations listed below apply to the design of vertical ladders and are not represented in the following figures or tables (Figs. 4.1 and 4.2):

- Permanent vertical ladders should be attached to a permanent structure,
- Located so as not to interfere with the opening and closing of hatches, grating, or other types of access,
- No impediments should intrude into the climbing space (for examples, electrical boxes, valves, actuators, or piping),
- If a work task requires the use of two hands, working from a vertical ladder is not appropriate. The work area should be provided with a work platform that provides a flat, stable standing surface. Refer to Fig. 4.3, "Landings (side mount)" and Fig. 4.4, "Vertical ladders to landings (ladder through platform)" for more details; and
- With the Administration's approval, "reasonable deviations" may be applied to facilitate this means of access. IACS UI (SC) 191 has interpreted this to be no more than 10% for vertical distances exceeding 6 m (19.5 ft).

Vertical Ladder Design

The following figures represent the different aspects of vertical ladders, their design and dimensioning:

- Figure 4.1, "Vertical ladders (general criteria),"
- Figure 4.2, "Staggered vertical ladder,"
- Figure 4.3, "Vertical ladders to landings (side mount),"
- Figure 4.4, "Vertical ladders to landings (ladder through platform)."

	Dimension	MSC.158(78)/UI SC 191 (PMA) requirement	PMA+ requirement
A	Distance between ladder attachments/securing devices	≤ 2.5 m (8.0 ft)	No additional requirement
B	Distance between ladder rungs (rungs evenly spaced throughout the full run of the ladder)	≥ 250 mm (10.0 in) and ≤ 350 mm (14.0 in)	≥ 275 mm (11.0 in) and ≤ 300 mm (12.0 in)
C	Distance between ladder stringers	≥ 350 mm (14.0 in)	400 to 450 mm (15.5 to 17.5 in)
D+	Ladder height (ladders over 6 m (19.5 ft) require intermediate/linking platforms)	≤ 6.0 m (19.5 ft)	No additional requirement
E	Stringer design	No specific requirement	Circular pipe with a diameter of 40 mm (1.5 in)
F	Rung design – (can be round or square bar; where square bar is fitted, orientation should be edge up)	Square bar 22 mm (0.9 in) × 22 mm (0.9 in)	Square bar 25 mm (1.0 in) × 25 mm (1.0 in) Round bar 25 mm (1.0 in) diameter
G	Ladder distance from surface (at 90 degrees)	≥ 150 mm (6.0 in)	Minimum 200 mm (8.0 in)
H	Horizontal Clearance (from ladder face and obstacles)	≥ 600 mm (23.5 in)	≥ 750 mm (29.5 in) or ≥ 600 mm (23.5 in) (in way of openings)
I	Overhead clearance	2.5m (8.0 ft)++	No additional requirement
J	Distance between ladder's centreline to any object that must be reached by personnel	No specific requirement	≤ 965 mm (38.0 in)
θ	Ladder angle of inclination from the horizontal	70-90 degrees	90 degrees
φ	Skew angle	≤ 2 degrees	0 degrees

+ = MSC.158(78) Table 2 requirement 1.10 allows for a single vertical ladder over 6 m (19.5 ft) in length for the inspection of the hold side frames in a single skin construction.

Fig. 4.1 Vertical ladders (general criteria)

4.2 Vertical Ladders

++ = The vertical distance of the uppermost section of the vertical ladder may be between 1.6 and 3 m (63 to 118 in), measured clear of the overhead obstructions in way of the tank entrance, if the ladder lands on a longitudinal or athwartship permanent means of access fitted within that range.

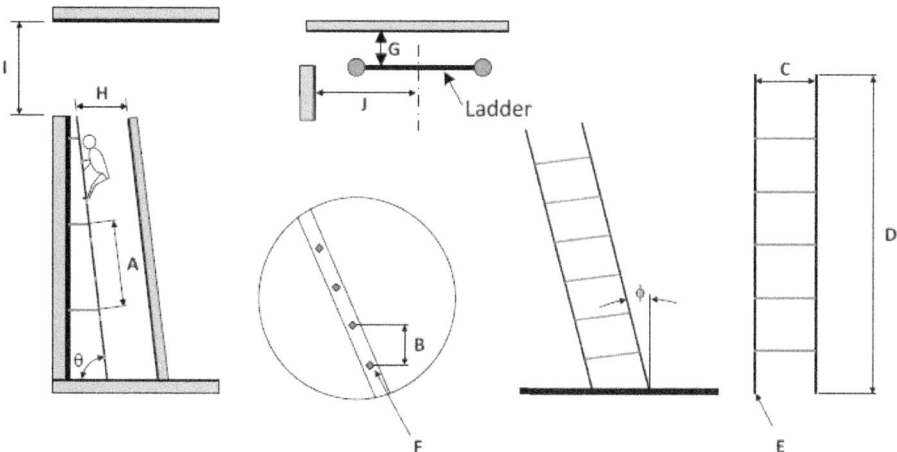

Fig. 4.1 (continued)

Climber Safety Devices

The use of climber safety devices is a standard safety and ergonomics practice on vertical ladders ≥4.5 m (15.0 ft). Cages should be used on vertical ladders over 4.5 m (15.0 ft) in height. Climber safety rails or cables should be used on vertical ladders in excess of 6.0 m (19.5 ft) whether cages are used or not since they are considered positive fall protection devices while cages are not. However, in tanks and holds, climber safety cages could be damaged by cargo being loaded, transported, and unloaded. Safety rails and cables may also be susceptible to damage but not as likely or severely as cages since the rails and cables fit tight against the ladder rungs. As a result, **PMA** and **PMA+** requirements do not require the use of climber safety devices or safety gates, but their use should be considered.

Note: This is a departure from safety and ergonomics practices but is acceptable for means of access for inspection in cargo tanks and holds. Additionally, consideration should be given to where such devices will prove helpful depending on the nature of the cargo and the device's potential susceptibility to damage or deterioration.

Climber Safety Cages

The use of climber safety devices is not a **PMA+** requirement. It is optional but strongly encouraged. Guidance for the construction of the safety cage is shown in Fig. 4.5, "Arrangement for cage of vertical ladder" and Fig. 4.6. "Cage of vertical ladder—side view." Cages should extend 1400 mm (55 in) above the top-landing surface. Cages equipped

	Dimension	MSC.158(78)/UI SC 191 (PMA) requirement	PMA+ requirement
A	Stringer width	≥ 350 mm (14.0 in)	400 to 450 mm (15.5 to 17.5 in)
B	Horizontal separation between two vertical ladders, stringer to stringer	≥ 200 mm (8.0 in)	≥ 200 mm (8.0 in) ≤ 460 mm (18.0 in)
C	Distance between ladder rungs (rungs evenly spaced throughout the full run of the ladder)	≥ 250 mm (10.0 in) and ≤ 350 mm (14.0 in)	≥ 275 mm (11.0 in) and ≤ 300 mm (12.0 in)
D*	Stringer height above landing or intermediate platform	≥ 1500 mm (59.0 in)	≥ 1500 mm (59.0 in)
E+	Rung design – (can be round or square bar; where square bar is fitted, orientation should be edge up)	Square bar 22 mm (0.9 in) × 22 mm (0.9 in)	Square bar 25 mm (1.0 in) × 25 mm (1.0 in) Round bar 25 mm (1.0 in) diameter
F	Horizontal separation between ladder and platform	≥100 mm (4.0 in) and ≤ 300 mm (12.0 in)	≥ 150 mm (6.0 in) and ≤ 300 mm (12.0 in)

Fig. 4.2 Staggered vertical ladder **a** side mount **b** ladder through the linking platform

with intermediate landings should extend 1400 mm (55 in) above the intermediate landing with the cage open on the side facing the landing. Consideration should be given to providing safety cages for ladders ≤4.5 m (15.0 ft) in height where a fall to a level or deck below the ladder base is possible (e.g., within 1825 mm (72 in) of the edge of a deck).

Climber Safety Rails or Cables

Listed below is guidance related to the use of climber safety rails or cables, where provided:

- For vertical ladders over 6.0 m (19.5 ft), a climber safety rail or cable should be considered, whether or not a safety cage is provided,
- Climber safety rail should be stainless steel flat bar and equipped with two safety slides, which can be attached to the flat bar or cable,

4.2 Vertical Ladders

G	Landing or intermediate platform width	Refer to Chapter 4, Figure 3 or 4 ±	
H	Stringer construction	No specific requirement	Round bar 40 mm (1.5 in) in diameter
I	Platform ladder to Platform ledge	No specific requirement	≥ 75 mm (3.0 in) and ≤ 150 mm (6.0 in)

+ = For **PMA+**, there should be a rung on the vertical ladder at the same height as the standing surface of the intermediate platform.

± = Refer to Figure 4-3, "Vertical ladders to landings (side mount)" or Figure 4-4, "Vertical ladders to landings (ladder through the platform)."

* = (1 November 2016) The minimum height of the handrail of resting platform is of 1,000 mm (39.5 in) [Technical Provision, resolution MSC.158(78), paragraph 3.3].

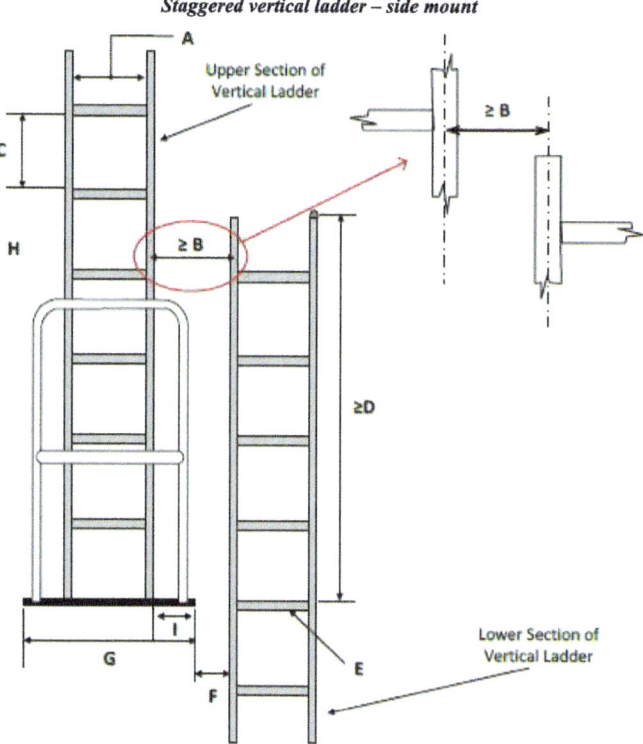

Staggered vertical ladder – side mount

Fig. 4.2 (continued)

Staggered vertical ladder – ladder through the linking platform

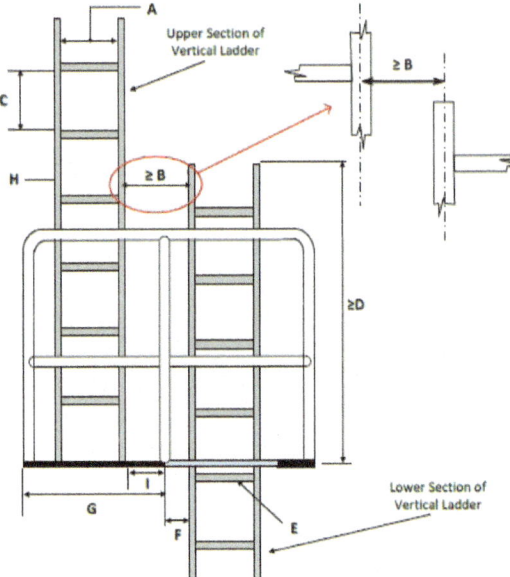

Fig. 4.2 (continued)

- Climber safety cables are recommended in place of rails in environments where any material may interfere with the rail itself; and
- If climber safety devices are used, the ladder stringers at a top landing should be designed to allow personnel to access any associate landing without unfastening (refer to Fig. 4.7, "Ladders with climber safety rails or cables").

Gates
Consideration should be given to the use of closable or self-closing gates on vertical ladders to prevent falls through ladder stringers. Chains or wire ropes do not provide the same level of safety as a gate.

Safety Drop Bars
All fixed ladders serving elevations 760 mm (30.0 in), or more above ground, platform or floor level should be equipped with drop bars or safety gates. Drop bars should be attached as follows:

(1) Side access ladders should hinge at the ladder side,
(2) Front access ladders should hinge at the right when facing the ladder from the platform side,

4.2 Vertical Ladders

	Dimension	MSC.158(78)/UI SC 191 (PMA) requirement	PMA+ requirement
A	Platform depth	Adequate dimensions	≥ 750 mm (29.5 in)
B	Platform width	Adequate dimensions	≥ 925 mm (36.5 in)
C	Ladder distance from surface	≥ 150 mm (6.0 in)	≥ 200 mm (8.0 in)
D	Horizontal separation between ladder and platform	≥ 100 mm (4.0 in) and ≤ 300 mm (12.0 in)	≥ 150 mm (6.0 in) and ≤ 300 mm (12.0 in)

+ = Other vertical ladder measurements apply.

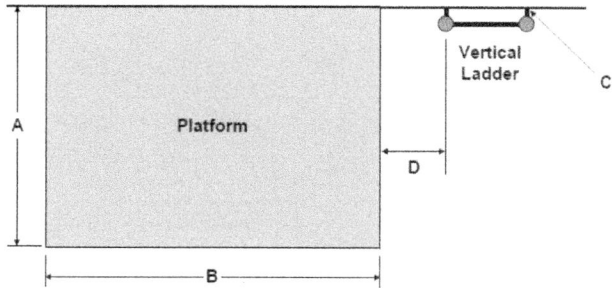

Fig. 4.3 Vertical ladders to landings (side mount)+

(3) Drop bars should not be placed beyond the outer edge of the platform; and
(4) Chains should not be used in lieu of a drop bar.

Safety Gates

Where a self-closing safety gate is provided, the following should apply:

(1) The self-closing safety gate should be installed at the top of each ladder and should cover the full width of the opening between the ladder stringers,
(2) The gate should open away from the person climbing up the ladder,
(3) Safety gates should be sufficiently robust to resist the full weight of a 90 kg (200 lbs) person in both the vertical and horizontal direction; and
(4) Chains should not be used in lieu of a safety gate.

	Dimension	MSC.158(78)/UI SC 191 (PMA) requirement	PMA+ requirement
A	Vertical ladder opening	≥ 600 mm (23.5 in)	≥ 750 mm (29.5 in)
B	Distance from front of vertical ladder to back of platform opening	≥ 600 mm (23.5 in)	≥ 750 mm (29.5 in)
C &D	Minimum clear standing area in front of ladder opening	No specific requirement	≥ 750 mm (29.5 in) × ≥ 925 mm (36.5 in)
E	Additional platform width for intermediate landing	No specific requirement	≥ 925 mm (36.5 in)
F	Horizontal separation between ladder and platform	≥ 100 mm (4.0 in) and ≤ 300 mm (12.0 in)	≥ 150 mm (6.0 in) and ≤ 300 mm (12.0 in)

+ = Other vertical ladder measurements apply.

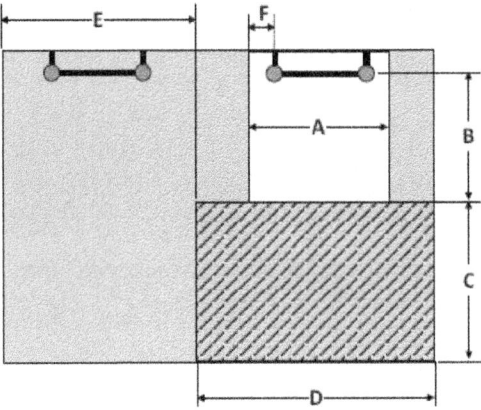

Fig. 4.4 Vertical ladders to landings (ladder through platform)+

Fall Protection from Secondary Fall Points

General

The use of fall protection from secondary fall points is a standard safety and ergonomics practice on vertical ladders. However, in tanks and holds, this additional fall protection could be damaged by cargo being loaded, transported, and unloaded. As a result, **PMA** and **PMA+** requirements do not require the use of fall protection from secondary fall points, but their use should be considered.

4.2 Vertical Ladders

	Dimension	Measurement
A	Distance from centreline of ladder rung to point of radius of the safety cage horizontal guards	350 mm (14.0 in)
B	Horizontal guard radius	Horizontal guard at bottom of cage – 425 mm (16.5 in)
		All other horizontal guards – 350 mm (14.0 in)
C	Vertical separation of horizontal guard placement	≤ 1200 mm (47.0 in)

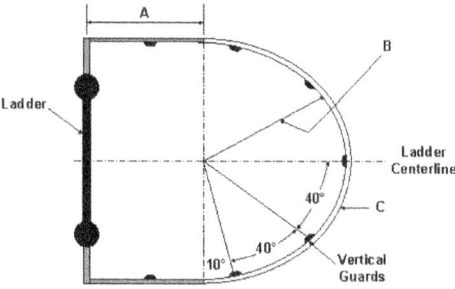

Fig. 4.5 Arrangement for cage of vertical ladder

Note: This is a departure from safety and ergonomics practices but is acceptable for means of access for inspection in cargo tanks and holds. Additionally, consideration should be given to where such fall protection will prove helpful depending on the nature of the cargo and the guardrail's potential susceptibility to damage or deterioration.

Vertical ladders should not be located within 1.83 m (6 ft) of other nearby potential fall points (including the deck edge, cargo holds and lower decks) without additional fall protection such as guardrails. Additional fall protection should be provided for the ladder climber for the case:

(1) If a vertical ladder (of any height) is located within 1.83 m (6 ft) of another and nearby potential (secondary) fall point (for example overboard or to a lower deck or landing).
    and
(2) If the potential fall distance is greater than 4.6 m (15 ft). (The potential fall distance is the height of the ladder plus the height of the secondary fall).
    and
(3) If no active fall protection is fitted to the ladder (a safety cage is not considered to provide active protection).

Dimension		Measurement
A	Distance above standing surface	≥ 2,200 mm (86.5 in)
		≤ 2,500 mm (98.5 in)
B	Vertical separation of horizontal guard placement	≥ 1,140 mm (45.0 in)
		≤ 1,220 mm (48.0 in)

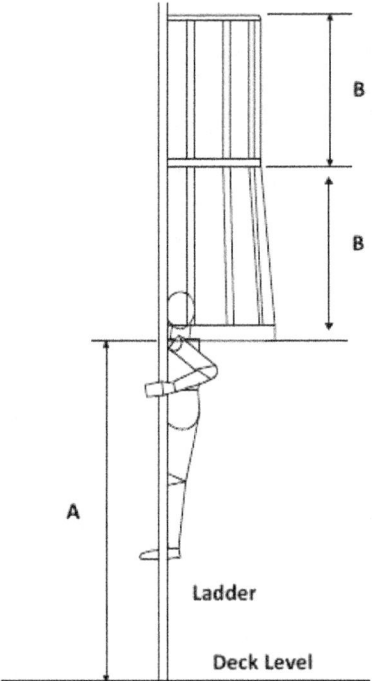

Fig. 4.6 Cage of vertical ladder—side view

then

(4) Additional fall protection to the ladder climber should be provided, regardless of whether a climber safety cage is fitted to the ladder, as described in the section "Protection for Vertical Ladders Without Safety Cages or Climber Safety Rails/Cables" below.

Note: Passive fall protection is a safety design or device that requires a person to take no specific action prior to a potential loss, for example, a safety cage permanently fitted to a ladder. Active fall protection is a safety design or device that actively (or directly) requires

4.2 Vertical Ladders

	Dimension	Measurement
A	Distance to bottom of climber safety rail	≥ 900 mm (35.5 in) ≤ 950 mm (37.5 in)
B	Inside clearance	≥ 225 mm (9.0 in) ≤ 250 mm (10.0 in)
C	Inside clearance	380 mm (15.0 in)
D	Top of climber safety rail and handrail height above upper standing surface	≥ 1,070 mm (42.0 in)
E	Climber safety rail height above upper standing surface	1,070 mm (42.0 in)
F	Distance from upper standing surface	200 mm (8.0 in)

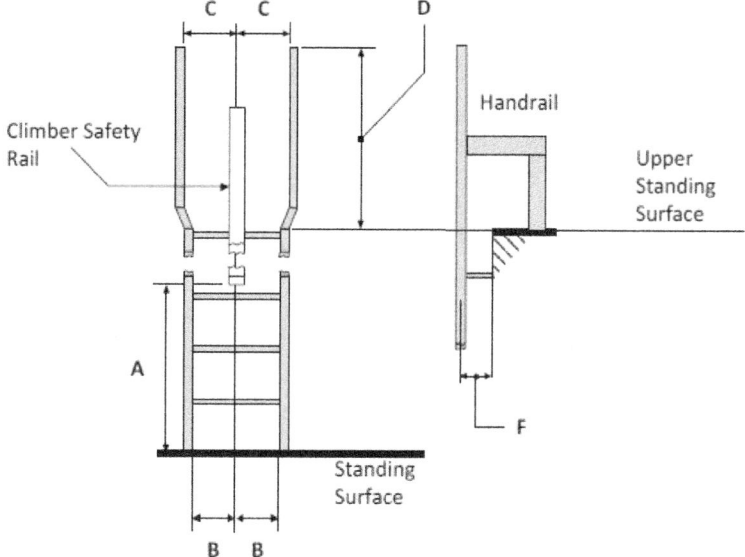

Fig. 4.7 Ladders with climber safety rails or cables

a person to take specific actions before a potential loss, for example, donning a fall arrestor fitted to both the ladder and the climber.

Table 4.2 Guardrail requirements for vertical ladders without safety cages or climber safety rails/cables

Dimension	Measurement
Height of vertical guardrail	Height should extend to within 760 mm (30.0 in) of the top of the ladder
Width of vertical guardrail	Protection should be provided for a minimum of 1220 mm (48.0 in) on each side of the centreline of the ladder, space permitting
Distance between guardrail courses or tiers	Maximum of 460 mm (18.0 in) should be provided between guardrail courses or tiers of the guardrail extension. The measurement should be taken from the course or tier's outside diameter to outside diameter as shown in Fig. 4.8

Protection for Vertical Ladders Without Safety Cages or Climber Safety Rails/Cables

The following should apply to vertical ladders less than of 4.5 m (15.0 ft) that are not fitted with a safety cage or a safety rail. Guardrail requirements are found in the following:

(1) Table 4.2, "Guardrail requirements for vertical ladders without safety cages or climber safety rails/cables" provides the dimensions for requirements to guardrails near the base of the ladder; and
(2) Fig. 4.8, "Front view of guardrail requirements for vertical ladders without safety cages or climber safety rails/cables" and Fig. 4.9, "Side view of guardrail requirements for vertical ladders with or without safety cages or climber safety rails/cables" provide dimensions and graphical representations (Fig. 4.10).

Other designs and arrangements that serve to protect personnel from falls may also be deemed to be acceptable (Table 4.3).

Individual Rung Ladders

The considerations listed below are applicable to the **PMA+** notation only and apply to individual rung ladders and are not represented in the following figure (Fig. 4.11):

- Individual rungs may be attached directly to a bulkhead, tank or steel structure and used as a vertical ladder, but should be limited to changes in vertical elevation of 3.6 m (12.0 ft) or less,
- Circular (round bar) rungs are preferred. Each rung should be attached to the structure in a manner that fully supports a climber and any design loads,

4.2 Vertical Ladders

	Dimension	Measurement
A	Horizontal spacing between ladder centreline and rail end	≥ 900 mm (35.5 in) ≤ 950 mm (37.5 in)
B	Inside clearance	≥ 225 mm (9.0 in) ≤ 250 mm (10.0 in)
C	Inside clearance	380 mm (15.0 in)
D	Top of climber safety rail and handrail height above upper standing surface	≥ 1,070 mm (42.0 in)
E	Climber safety rail height above upper standing surface	1,070 mm (42.0 in)
F	Distance from upper standing surface	200 mm (8.0 in)

* *Note: vertical ladder requirements apply (refer to the section on "Vertical ladders")*

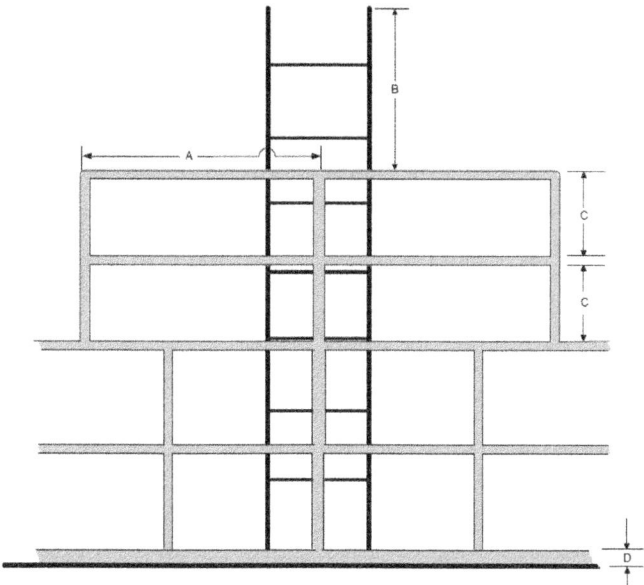

Fig. 4.8 Front view of guardrail requirements for vertical ladders without safety cages or climber safety rails/cables

Dimension		Measurement
A	Horizontal distance between ladder and rails	≤ 1,830 mm (72.0 in)
B	Vertical distance from top of rail to top of ladder	≤ 760 mm (30.0 in)

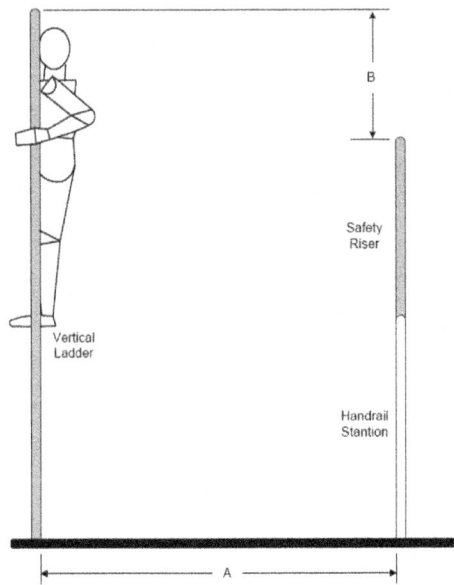

Fig. 4.9 Side view of guardrail requirements for vertical ladders without safety cages or climber safety rails/cables

- Rungs should be provided with lateral support for the foot; and
- It is recommended that square bar rungs should also be carried through the side stringers and attached by double continuous welding. Any other design must be in accordance to an international or national standard.

4.3 Inclined Ladders

This Subsection contains detailed guidance on the design of inclined ladders. The considerations listed below apply to the design of inclined ladders and are not represented in the following figures or tables.

4.3 Inclined Ladders

	Dimension	Measurement
A	Distance to bottom of climber safety rail	≥ 900 mm (35.5 in) ≤ 950 mm (37.5 in)
B	Inside clearance	≥ 225 mm (9.0 in)
		≤ 250 mm (10.0 in)
C	Inside clearance	380 mm (15.0 in)
D	Top of climber safety rail and handrail height above upper standing surface	≥ 1,070 mm (42.0 in)
E	Climber safety rail height above upper standing surface	1,070 mm (42.0 in)
F	Distance from upper standing surface	200 mm (8.0 in)

Note: assumes that the potential fall distance is greater than 4,575 mm (15 ft) where a safety cage, but no climber safety rail or cable, is present.

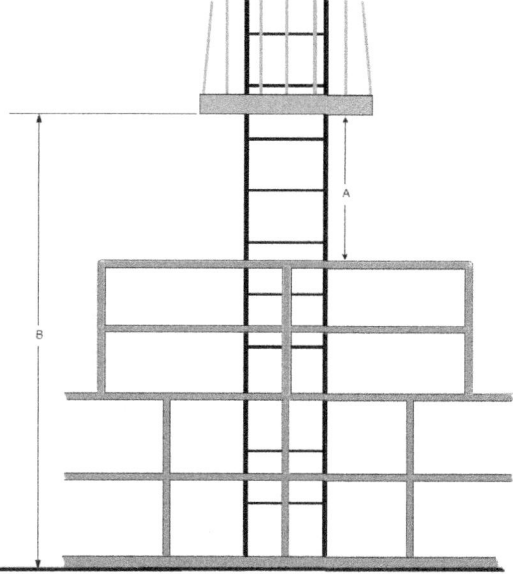

Fig. 4.10 Front view of guardrail requirements for vertical ladders with safety Cages and without climber safety rails/cables*

Table 4.3 Guardrail requirements for vertical ladders with safety cages and without climber safety rails/cables

Dimension	Measurement
Height of vertical guardrail	The height should extend to within 760 mm (30.0 in) of the lower edge of the safety cage
Width of vertical guardrail	Protection should be provided for a minimum of 1220 mm (48.0 in) on each side of the centreline of the ladder, space permitting
Distance between guardrail extension courses or tiers	A maximum of 460 mm (18.0 in) should be provided between guardrails courses or tiers of the guardrail extension. The measurement should be taken from the course or tier outside diameter to outside diameter as shown in Fig. 4.9
Horizontal distance between ladder and rails	1830 mm (72.0 in) (refer to Dimension A of Fig. 4.9, "Side view of guardrail requirements for vertical ladders without safety cages or climber safety rails/cables")

General

- Inclined ladders should be attached to a permanent structure,
- No impediments should intrude into the climbing space (for example, electrical boxes, valves, actuators, or piping),
- Inclined ladders and handrails should be located so as not to interfere with the opening and closing of hatches, gratings or manholes,
- Tread/steps should also be carried through the side stringers and attached by double continuous welding,
- IMO requires all inclined ladders should be provided with handrails of substantial construction on both sides; and
- Square handrails should be avoided.

Inclined Ladder Design

The following figures represent the different aspects of inclined ladders, their design and dimensioning:

- Figure 4.12, "Inclined ladders,"
- Figure 4.13, "Inclined ladders with landings;" and
- Figure 4.14, "Inclined ladder landing/platform."

4.4 Handles

	Dimension	Measurement
A	Rung width	≥ 400 mm (15.5 in) and ≤ 450 mm (17.5 in)
B	Rung depth – (can be round or square bar; where square bar is fitted, orientation should be edge up)	≥ 200 mm (8.0 in)
C	Rung design – (can be round or square bar; where square bar is fitted, orientation should be edge up)	Square bar 25 mm (1.0 in) × 25 mm (1.0 in) Round bar 25 mm (1.0 in) diameter
D	Distance between ladder rungs (rungs evenly spaced throughout the full run of the ladder)	≥ 275 mm (11.0 in) and ≤ 300 mm (12.0 in)
E	Height of foot slip protection	50 mm (2.0 in)

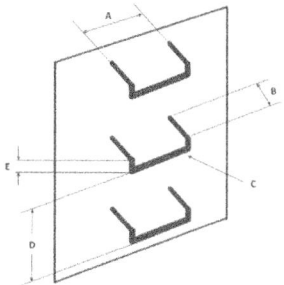

Fig. 4.11 Individual rung ladder design

Spiral Ladders

A spiral ladder is considered acceptable as an alternative for inclined ladders for Bulk Carriers. In this regard, the uppermost 2.5 m (8.0 ft) can continue to be comprised of the spiral ladder and need not change over to vertical ladders.

4.4 Handles

This section contains detailed guidance related to the design of handles. The considerations listed below apply to the design and placement of handles and are not represented in the following figures.

	Dimension	MSC.158(78)/UI SC 191 (PMA) requirement	PMA+ requirement
A	Handrail diameter	No specific requirement	≥ 40 mm (1.5 in) ≤ 50 mm (2.0 in)
B	Handrail height (from leading edge of tread)	≥ 890 mm (35.0 in)	≥ 915 mm (36.0 in) and ≤ 1,000 mm (39.5 in)
C	Tread/step spacing – equally spaced along entire ladder	≥ 200 mm (8.0 in) and ≤ 300 mm (12.0 in)	No additional requirement
D	Square bar step depth	No specific requirement	≥ 100 mm (4.0 in)
E	Handrail to handrail width	≥ 450 mm (17.5 in) for cargo holds ≥ 400 mm (15.5 in) for other areas	≥ 450 mm (17.5 in) and ≤ 560 mm (22.0 in)
F	Rung Design – (Can be round or square bar; where square bar is fitted, orientation should be edge up)	Square bar 22 mm (0.9 in) × 22 mm (0.9 in)	Square bar 25 mm (1.0 in) × 25 mm (1.0 in) Round bar 25 mm (1.0 in) diameter
G	Obstruction distance from the face of the inclined ladder	≥ 750 mm (29.5 in), except that in way of an opening this clearance can be reduced to 600 mm (23.5 in)	≥ 1,240 mm (49.0 in)
H	Vertical obstruction height above ladder	No specific requirement	≥ 2,130 mm (84.0 in)
I	Maximum continuous height	≤ 6 m (19.5 ft)	No additional requirement
J	Clearance between the handrail and a bulkhead or other obstruction	No specific requirement	≥ 75 mm (3.0 in)
θ	Angle of inclination	< 70 degrees	45-60 degrees

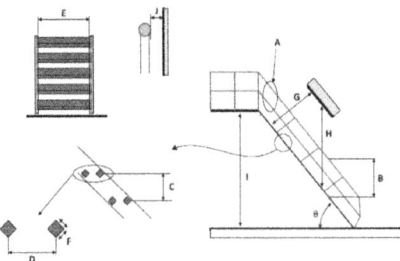

Fig. 4.12 Inclined ladders

4.4 Handles

	Dimension	MSC.158(78)/UI SC 191 (PMA) requirement	PMA+ requirement
A[+]	Clearance above ladder platforms	≥ 2.5 m (8.0 ft) and ≤ 3 m (10.0 ft)	No additional requirement
B & C	Maximum continuous height	≤ 6 m (19.5 ft)	No additional requirement
D	Deck to lower landing level	≥ 2.5 m (8.0 ft) and ≤ 6m (19.5 ft)	No additional requirement
E	Height of intermediate rail	500 mm (19.5 in)	535 mm (21.0 in)
F	Height of top rail	500 mm (19.5 in)	535 mm (21.0 in)
G	Landing/platform dimensions	Refer to Figure 4-14, "Inclined ladder landing/platform"	
H	Obstruction distance from the face of the inclined ladder	≥ 750 mm (29.5 in), except that in way of an opening this clearance can be reduced to 600 mm (23.5 in)	1,240 mm (49.0 in) (Minimum)
I	Vertical obstruction height above ladder	No specific requirement	2,130 mm (84.0 in)
J	Height of handrail	1,000 mm (39.5 in)	1,070 mm (42.0 in)
K	Stringer height above landing or intermediate platform	≥ 1,000 mm (39.5 in)	≥ 1,350 mm (53.0 in)
θ	Angle of inclination	< 70 degrees	45-60 degrees

[+] = The vertical distance of the uppermost section of the vertical ladder may be reduced to 1.6 m (63.0 in) to 3 m (118.0 in), measured clear of the overhead obstructions in way of the tank entrance, if the ladder lands on a longitudinal or athwartship permanent means of access fitted within that range.

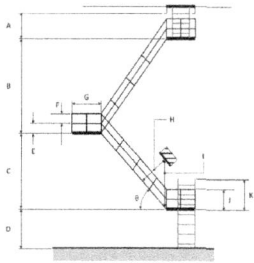

Fig. 4.13 Inclined ladders with landings

Dimension		MSC.158(78)/UI SC 191 (PMA) requirement	PMA+ requirement
A	Minimum landing width	Adequate dimensions	600 mm (23.5 in) for the last flight of inclined ladder. *Note: if landing is used to access another inclined ladder flight, then the landing width should be at least twice the width of the inclined ladder.*
B[+]	Minimum landing length	No specific requirement	975 mm (38.5 in)

[+] = Where inclined ladders change directions, it is recommended that intermediate landings along paths for evacuating personnel stretchers be 1,525 mm (60.0 in) in length to accommodate rotating the stretcher.

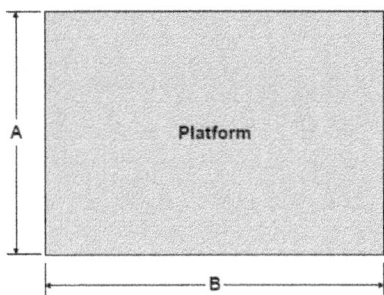

Fig. 4.14 Inclined ladder landing/platform

General

- Handles should be designed to accommodate personnel wearing either lightweight and medium weight gloves or cold weather gloves and mittens (refer to Fig. 4.15, "Handle dimensions"),
- Handles are particularly useful where a vertical ladder comes up to a manhole from the deck below where the ladder does not extend through the platform [refer to Fig. 4.16, "Handle Placement (Ladder not extending through platform)" or while passing through access hatches (refer to Fig. 4.17, "Handle placement (stepping through a vertical hatch)"],
- Handles should be accessible at all stages during climbing or traversing through access hatches (embarking and disembarking) and within reach of the shortest (e.g., 5th percentile female) user; and

4.4 Handles

	Dimension	MSC.158(78)/UI SC 191 (PMA) requirement	PMA+ requirement
A	Handle width	No specific requirement	≥ 300 mm (12.0 in) ≤ 350 mm (14.0 in)
B	Handle height	No specific requirement	100 mm (4.0 in)
C	Radius	No specific requirement	25 mm (1.0 in)
D	Round bar diameter	No specific requirement	25 mm (1.0 in)

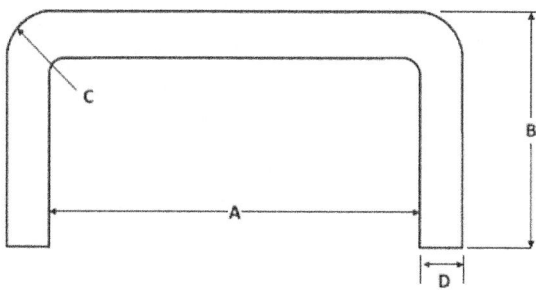

Fig. 4.15 Handle dimensions

- To provide for safe ascending and descending, while stepping onto or from ladders, individual ladder rungs or steps and through hatches or lightening holes, suitably located handles or hand grabs should be provided [refer to Fig. 4.18, "Handle placement (stepping to or from a vertical ladder)"].

Handle Design/placement

The following figures represent the different aspects of handle design, placement and dimensioning.

- Figure 4.15, "Handle dimensions,"
- Figure 4.16, "Handle placement (ladder not extending through platform),"
- Figure 4.17, "Handle placement (stepping through a vertical hatch);" and
- Figure 4.18, "Handle placement (stepping to or from a vertical ladder)."

	Dimension	MSC.158(78)/UI SC 191 (PMA) *requirement*	PMA+ *requirement*
	Four horizontal handles		
A	Handle height above top of ladder	No specific requirement	≥ 275 mm (11.0 in) and ≤ 300 mm (12.0 in)
E	Round Bar Diameter	No specific requirement	25 mm (1.0 in)
	Two vertical handles		
B	Height from top deck to handle	No specific requirement	200 mm (8.0 in)
C	Clearance between handles	No specific requirement	400 mm (15.5 in)
D	Height of handles	No specific requirement	1,000 mm (39.5 in)
E	Round bar diameter	No specific requirement	25 mm (1.0 in)

+ = Other vertical ladder measurements apply (refer to "Vertical ladders")

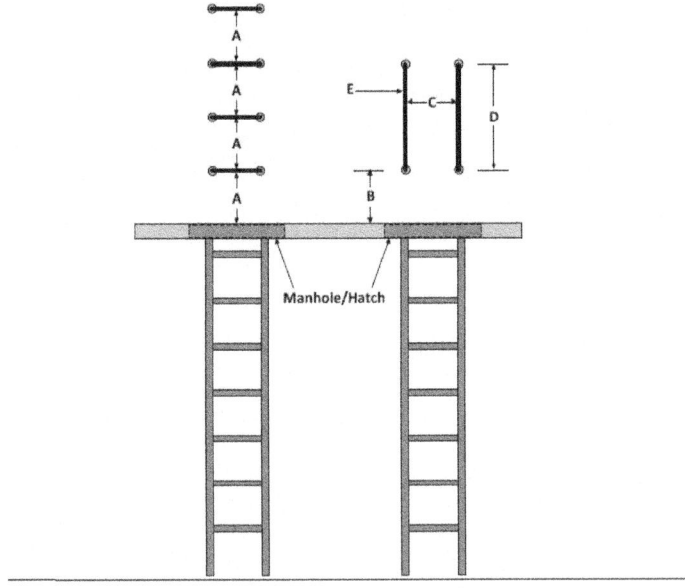

Fig. 4.16 Handle placement (ladder not extending through platform)[+]

4.4 Handles

	Dimension	MSC.158(78)/UI SC 191 (PMA) requirement	PMA+ requirement
A	Handle height (above top of opening)	No specific requirement	100 mm (4.0 in)
B	Distance between lower and upper portion of the hatch	≥ 800 mm (31.5 in)	≥ 1,000 mm (39.5 in)
C	Height required for a step	> 600 mm (23.5 in)	No additional requirement
D	Handle width	No specific requirement	≥ 300 mm (12.0 in)
E	Step height	No specific requirement	≥ 275 mm (11.0 in) and ≤ 300 mm (12.0 in)
F	Step width, hatch width	No specific requirement	≥ 800 mm (31.5 in) or ≥ Hatch width
G	Step depth (not shown in figure)	No specific requirement	≥ 275 mm (11.0 in) and ≤ 300 mm (12.0 in)
A	Handle height (above top of opening)	No specific requirement	100 mm (4.0 in)

+ = Handles and steps are placed on both sides of the hatch

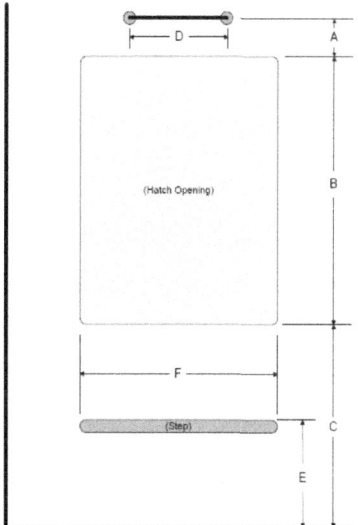

Fig. 4.17 Handle placement (stepping through a vertical hatch)[+]

	Dimension	MSC.158(78)/UI SC 191 (PMA) requirement	PMA+ requirement
A	Length of handle	No specific requirement	≥ 300 mm (12.0 in)
B	Handle height above landing or platform	No specific requirement	≥ 1,270 mm (50.0 in)
C	Ladder stringer height above platform	≥ 1,500 mm (59.0 in)	≥ 1,500 mm (59.0 in)
D	Horizontal separation between vertical ladders and platform	≥ 100 mm (4.0 in) ≤ 300 mm (12.0 in)	≥ 150 mm (6.0 in) and ≤ 300 mm (12.0 in)
E	Horizontal separation between vertical ladder and handle	No specific requirement	≥ 225 mm (9.0 in) ≤ 450 mm (17.5 in)

* *Note: vertical ladder requirements apply (refer to Chapter 4, section 2, "Vertical ladders")*

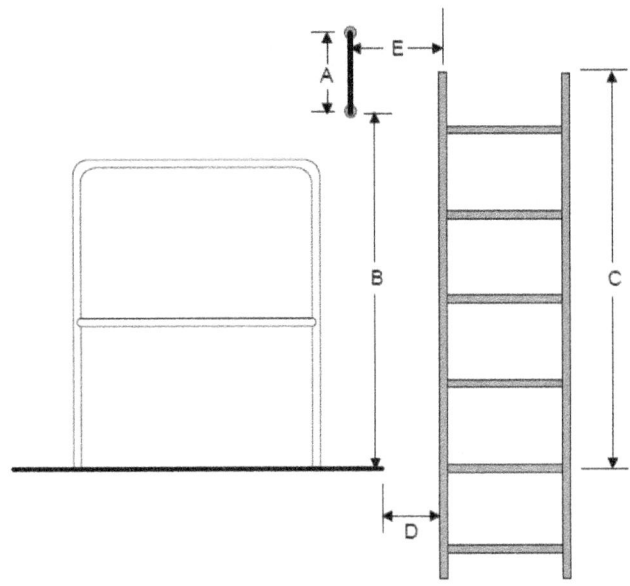

Fig. 4.18 Handle placement (stepping to or from a vertical ladder)

Hatches

5.1 General

This chapter contains guidance related to the design of hatches. The guidance included in the figures and tables below provides the design attributes (application) and the IMO (**PMA**) and **PMA+** dimension requirements. There are a few instances where IMO means of access requirements do not provide specific design dimensioning. In these instances, the **PMA+** dimensions may be used as guidance.

5.2 General Principles

The principles listed below apply to the design of hatches and lightening holes and are not represented in the following figures or tables.

- For access through horizontal hatches, the dimensions shall be sufficient to allow a person wearing a self-contained air-breathing apparatus and protective equipment to ascend or descend any ladder without obstruction and also provide a clear opening to facilitate the movement of an injured person through the hatch, and
- Where hatch covers are heavy [e.g., above 11 kg (24.3 lbs)] or unwieldy, aids shall be provided to assist in lifting or lowering the hatch cover.

	Dimension	MSC.158(78)/UI SC 191 (PMA) *requirement*	PMA+ *requirement*
A	Access – vertical height	≥ 800 mm (31.5 in)	≥ 1,000 mm (39.50 in)
B	Access – horizontal width	≥ 600 mm (23.5 in)	≥ 800 mm (31.50 in) preferred
C[+]	Height above deck or stepping tread	≤ 600 mm (23.5 in)	No additional requirement

[+] = If a vertical opening is at a height of more than 600 mm (23.5 in), steps and handgrips are to be provided. In such arrangements it is to be demonstrated that an injured person can be easily evacuated. For more guidance, refer to Chapter 4, section 4.2, Figure 4-17, "Handle placement (stepping through a vertical hatch)".

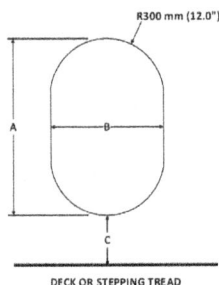

Fig. 5.1 Hatch design

5.3 Hatch Design

Figure 5.1, "Hatch design" and Fig. 5.2, "Hatch design (alternative arrangement)" represent IMO and IACS approved shapes and dimensioning and any **PMA+** requirements. Figure 5.1, "Hatch design" illustrates the dimensioning for access through vertical openings/manholes, in swash bulkheads, floors, girders and web frames providing passage through the length and breadth of the space. Figure 5.2, "Hatch design (alternative arrangement), illustrates an IACS-approved alternative design for access. This design is subject to the verification of easy evacuation of injured or stretcher borne personnel.

5.4 Horizontal Hatch Access Near a Coaming

For access through horizontal openings, hatches or manholes, the minimum clear opening shall not be less than 600 mm × 600 mm (23.5 in × 23.5 in). When access to a cargo hold is arranged through the cargo hatch, the top of the ladder shall be placed as close as

5.5 Horizontal Hatch Access Near a Coaming

	Dimension	MSC.158(78)/UI SC 191 (PMA) *requirement*	PMA+ *requirement*
A	Access – vertical height	≥ 850 mm (33.5 in)	≥ 1,000 mm (39.5 in)
B	Access – horizontal width	≥ 620 mm (24.5 in)	≥ 800 mm (31.5 in) preferred
C+	Height above deck or stepping tread	≤ 600 mm (23.5 in)	No additional requirement

+ = If a vertical opening is at a height of more than 600 mm, steps and handgrips are to be provided. In such arrangements it is to be demonstrated that an injured person can be easily evacuated. For more guidance refer to Chapter 4, section 2, Figure 4-17, "Handle placement (stepping through a vertical hatch)".

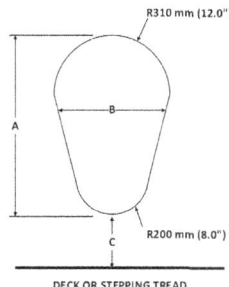

Fig. 5.2 Hatch design (alternative arrangement)

possible to the hatch coaming. Access hatch coamings having a height greater than 900 mm (35.5 in) shall also have steps on the outside in conjunction with the ladder (refer to Chapt. 4, section "Inclined Ladders", Fig. 5.4, "Access hatch heights of ≥900 mm (35.5 in)") (Fig. 5.3).

5.5 Horizontal Hatch Access Near a Coaming

For access through horizontal openings, hatches or manholes, the minimum clear opening shall not be less than 600 mm × 600 mm (23.5 in × 23.5 in). When access to a cargo hold is arranged through the cargo hatch, the top of the ladder shall be placed as close as possible to the hatch coaming. Access hatch coamings having a height greater than 900 mm (35.5 in) shall also have steps on the outside in conjunction with the ladder (refer to Fig. 5.4, "Access hatch heights of ≥900 mm (35.5 in)").

	Dimension	MSC.158(78)/UI SC 191 (PMA) *requirement*	PMA+ *requirement*
A	Distance from step to access hatch	600 mm (23.5 in)	No additional requirement
B	Step depth	No specific requirement	≥ 275 mm (11.0 in) and ≤ 300 mm (12.0 in)
C+	Step height	No specific requirement	Refer to footnote 300 mm (12.0 in) (maximum)
D	Height to require steps along with the ladder	≥ 900 mm (35.5 in)	No additional requirement
E	Dimension inside of hatch without obstruction	≥ 600 mm (23.5 in)	≥ 750 mm (29.5 in)
F	Distance from ladder to hatch coaming	As close as possible	≥ 200 mm (8.0 in)

+ = The limiting height is dimension 'A'. This height is set by the crotch height of the 5th percentile female. Thus, once 'D' exceeds 900 mm (35.5 in), a step is needed. Therefore, 'C' could be anything from 25 mm (1.0 in) up to 300 mm (12.0 in).

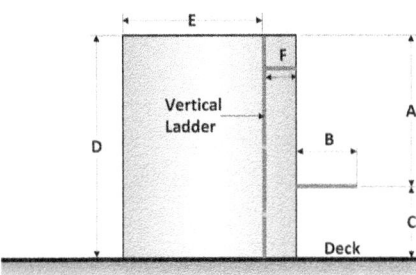

Fig. 5.3 Handle placement (stepping through a vertical hatch)

Horizontal Hatch Access Through a Deck (for PMA+ notation)

For access to the deck from a ladder below, the top of the ladder shall be placed within 50 mm (2.0 in) of the leading edge of the hatch opening (refer to Fig. 5.5, "Horizontal hatch access through a deck"). Minimum dimensions of the opening (round or square) are 810 mm (32.0 in).

5.5 Horizontal Hatch Access Near a Coaming

	Dimension	MSC.158(78)/UI SC 191 (PMA) *requirement*	PMA+ *requirement*
A	Distance from step to access hatch	600 mm (23.5 in)	No additional requirement
B	Step depth	No specific requirement	≥ 275 mm (11.0 in) and ≤ 300 mm (12.0 in)
C+	Step height	No specific requirement	Refer to footnote 300 mm (12.0 in) (maximum)
D	Height to require steps along with the ladder	≥ 900 mm (35.5 in)	No additional requirement
E	Dimension inside of hatch without obstruction	≥ 600 mm (23.5 in)	≥ 750 mm (29.5 in)
F	Distance from ladder to hatch coaming	As close as possible	≥ 200 mm (8.0 in)

+ = The limiting height is dimension 'A'. This height is set by the crotch height of the 5th percentile female. Thus, once 'D' exceeds 900 mm (35.5 in), a step is needed. Therefore, 'C' could be anything from 25 mm (1.0 in) up to 300 mm (12.0 in).

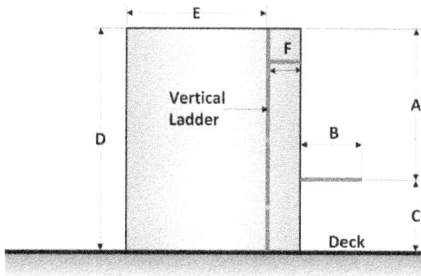

Fig. 5.4 Access hatch heights of ≥ 900 mm (35.5 in)

	Dimension	Requirement
A	Dimension of opening (circular or rectangular)	≥ 810 mm (32.0 in)
B	Ladder to edge of opening separation	≤ 50 mm (2.0 in)

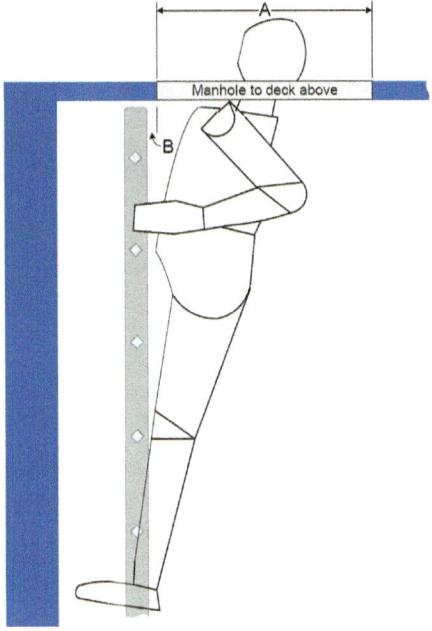

Fig. 5.5 Horizontal hatch access through a deck

Alternative Means of Access

6.1 General

This chapter provides guidance relating to the design of alternative means of access. IMO requirements, under certain circumstances, allow for the use of alternative means of access in place of permanent means of access. This chapter contains no **PMA** or **PMA+** (or equivalent) criteria.

Definitions

Alternative means of access: A common term for portable or movable means of access provided for survey in areas otherwise not accessible, these include, but are not limited to, such devices as:

- Hydraulic arm fitted with a stable base,
- Wire lift platform,
- Staging,
- Rafting,
- Robot arm or remotely operated vehicle (ROV),
- Portable ladders; and
- Other means of access, approved by and acceptable to Class.

Movable means of access: Devices like a cherry picker or other means, which are not normally kept onboard. When such means are provided as an alternative to the permanent means of access, they should be kept onboard and capable of being operated by the vessel's crew.

Portable means of access: Equipment that may be hand carried by the crew (e.g., ladders, small platforms and rafts).

6.2 Application of SOLAS and MSC Regulations

SOLAS REG.II-1/3-6, "Access to and within spaces in, and forward of the cargo area of oil tankers and bulk carriers," allows for the use of alternative means of access to areas requiring inspection. Listed below are several excerpts:

- From SOLAS regulation II-1/3-6, 2.2: "Where a permanent means of access may be susceptible to damage during normal cargo loading and unloading operations or where it is impracticable to fit permanent means of access, the Flag State Administration may allow, in lieu thereof, the provision of movable or portable means of access, as specified in the Technical Provisions, provided that the means of attaching, rigging, suspending or supporting the portable means of access forms a permanent part of the ship's structure."
- From MSC.158(78), Table 1: "Means of access for oil tankers, resolution, 1.2 For tanks of which the height is less than 6 m, alternative means of as defined in paragraph 3.9 of the Technical Provisions or portable means may be utilised in lieu of the permanent means of access."

6.3 IMO Requirements for Alternative Means of Access

The following documents were used as references and provide details about IMO requirements for alternative means of access and should be used for reference:

- IACS Recommendation No. 39—Safe use of rafts or boats for survey,
- IACS Recommendation No. 42—Guidelines for use of remote survey techniques,
- IACS Recommendation No. 72—Confined space safe practice,
- IACS Recommendation No. 78—Safe use of portable ladders for close-up surveys,
- IACS Recommendation No. 90—Ship structure access manual,
- IACS Recommendation No. 91—Guidance for approval/acceptance of alternative means of access,
- IACS Unified Requirement—Z10.1 Hull surveys of oil tankers,
- IACS Unified Requirement—Z10.2 Hull surveys of bulk carriers,
- IACS Unified Requirement—Z10.4 Hull surveys of double hull oil tankers; and
- IACS Unified Requirement—Z10.5 Hull surveys of double skin bulk carriers.

Note: IACS Recommendations are occasionally updated. It is the responsibility of the reader to check and see if there have been any updates since the publication of this document. If so, the updated Recommendations should be used in conjunction with this text. The most current IACS Recommendations can be obtained at the following web site: http://www.iacs.org.uk.

6.4 Guidance for Alternative Means of Access

This section provides guidance for use of the alternative means of access requirements contained in MSC.158(78) "Amendments to the Technical Provisions for means of access for inspections."

Portable Ladders

Portable ladders may be used for access to structural members as supplementary and/or additional to permanent means of access in accordance with SOLAS II-1/3-6 and should be included in the Ship Structure Access Manual. Also, the requirements of IACS Recommendation No. 78 "Safe use of portable ladders for close-up surveys" should be used when specified for use in the Ship Structure Safe Access Manual as a portable means of access.

General Guidance

- The asset owner should verify that equipment selected for temporary work affords adequate protection against the risks of falls from a height,
- A freestanding portable ladder with a maximum length of 5 m (16.4 ft) may be used for infrequent inspections,
- Ladders should not be tied or fastened together to create longer sections,
- Step ladders, hanging ladders and portable ladders more than 5 m (16.4 ft) long should only be utilised if fitted with a mechanical device to secure the upper end of the ladder. A mechanical device such as hooks for securing ladder at the upper end is considered an appropriate securing device if capable of preventing movement fore/aft and sideways,
- Ladders should not be loaded beyond their maximum intended load or beyond the manufacturer's rated capacity; and
- The feet of portable ladders should be prevented from slipping during use by securing the ladder stringers (stiles) at or near their upper and lower ends, by an anti-slip device or by other arrangements of equivalent effectiveness. Slip resistant feet alone should

not be used as a substitute for the care in placing or otherwise securing a ladder upon a slippery surface.

Non-self-supporting ladders should be used at an angle where the horizontal distance from the top support to the foot of the ladder is approximately one-quarter of the working length of the ladder. Ladder rungs, cleats and steps should be parallel, level and uniformly spaced when the ladder is in position for use. Portable ladders should be used on top of bottom or deep stringer platform so that free falling height does not exceed 6 m (19.5 ft). If it is necessary to exceed this height, there should be at least 3 m (10.0 ft) of water above the highest structural element in the bottom to provide a "cushion" or safety harnesses may be used. The freefalling height above the water should not exceed 6 m (19.5 ft). The rungs and steps of portable ladders should be designed to minimise slipping (e.g., corrugated, knurled, dimpled and/or coated with skid resistance material, etc.). Ladder components and surfaces should be smooth to prevent snagging of clothing and injury from punctures or lacerations. Self-supporting and non-self-supporting portable ladders should support at least four times the maximum intended load.

Detailed Guidance
When portable ladders are used for access to an upper landing surface, the ladder stringers (stiles) should extend at least 900 mm (36.0 in) above the upper landing surface. When such an extension is not possible, the ladder must be secured and a grasping device such as a grab rail or handle should be provided to assist workers in mounting and dismounting the ladder. A ladder extension should not deflect under a load that would cause the ladder to slip off its supports.

Rungs, cleats and steps of portable ladders should not be spaced less than 275 mm (11.0 in) apart, nor more than 300 mm (12.0 in) apart, along the ladder's stringers (stiles).

Rungs, cleats and steps at the base section of extension trestle ladders should not be less than 200 mm (8.0 in) nor more than 450 mm (17.5 in) apart, between centrelines of the rungs, cleats or steps. The rung spacing on the extension section should not be less than 150 mm (6.0 in) or greater than 300 mm (12.0 in).

The minimum clear distance between stringers (stiles) for all portable ladders must be at least 300 mm (12.0 in) for ladders 3 m (10.0 ft) or less in overall length and should increase at least 6.5 mm (0.25 in) for each additional 600 mm (23.5 in) of ladder length.

Operational Considerations
All ladders should be maintained free of oil, grease and other slipping hazards. All ladders should be used only for their designed purpose, on stable and level surfaces unless secured to prevent accidental movement. The manner in which portable ladders can most safely be used by workers should be specified and addressed by the Ship's Safety Management System. Aluminium ladders may be used in cargo tanks but cannot be stored in the cargo area or other gas dangerous spaces. Portable ladders should rest on a stable,

6.4 Guidance for Alternative Means of Access

strong, suitably sized, immobile footing so that the rungs remain horizontal. Suspended ladders should be attached in a manner so that they cannot be displaced and so that swinging is prevented. Suspended ladders should be attached in a manner so that they cannot be displaced, and swinging is prevented. Areas around the top and bottom of ladders should be kept clear and clean. Ladders should not be moved, shifted or extended while in use. Personnel should face the ladder when moving up or down with at least one hand to grasp the ladder when climbing. Carrying objects or loads that could cause loss of balance and falling should be avoided. When climbing ladders in tanks containing water, the surveying personnel should wear "flotation" aids. A flotation aid is a simple form of lifejacket, which does not impede climbing, or a self-inflatable lifejacket.

Ladders should not be used on slippery surfaces unless secured or provided with slip-resistant feet to prevent accidental movement. Slip resistant feet alone should not be used as the only slip prevention technique. Neither should they be a substitute for the care in placing or otherwise securing a ladder upon a slippery surface. The use of ladders with broken or missing rungs or steps, broken or split stringers (stiles), or other faulty or defective construction is prohibited. When ladders with such defects are discovered, they should be immediately withdrawn from service. Inspection of metal ladders should include checking for corrosion of interiors of open end, hollow rungs. All ladders should be inspected prior to use.

Hydraulic Arm Vehicles

Hydraulic arm vehicles or aerial lifts ("cherry pickers") may be used to enable the examination of the cargo hold structure on bulk carriers not accessible by permanent ladders. In the Ship Structural Access Manual, the Cherry Pickers may be accepted as movable means of access for use up to 17 m (56.0 ft) above the tank top.

General Guidance

Asset owners are responsible for verifying that moveable means of access are suitable for the intended uses on the vessel. Qualified personnel should operate the vehicle and there should be proof that the vehicle has been properly maintained, at least to manufacturer's requirements. Lift controls, including safety devices should be serviceable and should be operated throughout the range prior to use. Operators should be trained. Lift controls, including safety devices should be tested daily. Permissible load and reach limitations should be understood and not exceeded. Operators should work from within the basket. The standing platform should be fitted with anchor points for attaching fall arrest systems. Body belts (such as harnesses) with lanyards should be used. Raising and lowering controls are required and labelled accordingly. Lowering controls should override the raising controls. For more detail regarding the labelling of controls, refer to Section 8, "Labelling, Signs, Graphics and Symbols" of *Applying Physical Ergonomics to Modern Ship*

Design (Olsen; Springer: 2024). Whenever internal combustion engine powered equipment exhausts in enclosed spaces, tests should be made and recorded to see that personnel are not exposed to unsafe concentrations of toxic gases or oxygen deficient atmospheres. Belts, gears, shafts, pulleys, sprockets, spindles, drums, fly wheels, chains or other reciprocating, rotating or other moving parts or equipment should be guarded if such parts are exposed to contact by personnel, or otherwise create a hazard.

Operational Considerations

Unless designed otherwise, aerial lift vehicles should not be moved when the boom is elevated in a working position with personnel in the basket. For those vehicles equipped with a self-levelling platform, care should be taken that the locking device is engaged after vehicle manoeuvring to verify that the platform is fixed. Brakes should be set, outriggers used (if so equipped) and wheels chocked (if on an incline). Potential crushing hazards (e.g., booming into the overhead, pinch point) should be avoided. Personal flotation devices (PFD) should be used when working over water. The Ship's Safety Management System should address the operation and training in the use of this type of equipment.

Wire Lift Platform

Wire lift platforms may be used for inspection of structural members of ballast tanks, cargo oil tanks and cargo hold. Such equipment should be rated for more than one person and be operated by suitable authorised personnel. If carried onboard and included in the Ship Structure Access Manual, designers will have to take into consideration safety aspects associated with deployment and use of such means and access. The platform and equipment, including fixed points to the vessel's structure, should be approved on behalf of the Flag State Administration being based on a recognised international or national standard. Approval of wire lift platforms should address the following:

- Accidental loss of balance,
- Permissible weight,
- Protection against overload,
- Secondary means of escape,
- Guardrails,
- Permissible loads,
- Permanent marking of loads; and
- Recovery in the event of power loss.

General Guidance

A qualified engineer or a qualified person competent in structural design should design the personnel platform and suspension system. The suspension system should be designed to prevent tipping of the platform due to movement of employees occupying the platform. The personnel platform should be conspicuously posted with a plate or other permanent marking which indicates the weight of the platform, and its rated load capacity. Means should be provided for using fall protection with lifelines tended above the platform. Whenever internal combustion engine powered equipment exhausts in enclosed spaces, tests should be made and recorded to see that personnel are not exposed to unsafe concentrations of toxic gases or oxygen deficient atmospheres. Belts, gears, shafts, pulleys, sprockets, spindles, drums, fly wheels, chains or other reciprocating, rotating, or other moving parts or equipment should be guarded if such parts are exposed to contact by personnel, or otherwise create a hazard.

Detailed Guidance

Each personnel platform should be equipped with a guardrail system designed as depicted in Fig. 3.2 "Walkway and ramp design." Access gates, if installed, should not swing outward during hoisting. Load lines should be capable of supporting at least seven (7) times the maximum intended load, except that where rotation resistant rope is used, the lines should be capable of supporting at least ten (10) times the maximum intended load.

Operational Considerations

Hoisting of the personnel platform should be performed in a slow, controlled, cautious manner with no sudden movements. Rigging of wires should be in accordance with manufacturer's recommendations and conducted by qualified personnel. Fix points to which the wires will be connected should be examined before each use and verified in good condition (free of wastage, fractures, etc.). Wire rope should be taken out of service when any of the following conditions exist:

- In running ropes, six randomly distributed broken wires in one lay or three broken wires in one strand in one lay,
- Wear of one-third the original diameter of outside individual wires,
- Kinking, crushing, bird caging, or any other damage resulting in distortion of the rope structure; and
- Evidence of any heat damage from any cause.

Portable Platforms

Portable platforms may be used as a portable means of access, provided that the platform and equipment, including fixed points to the vessel's structure are specifically designed for the task and approved on behalf of the Flag State Administration based on a recognised international or national standard. Portable platforms not more than 3 m (10 ft) in length may be used for access between longitudinal permanent means of access and the structural member to be accessed. Guardrails (see Sect. 6.3, Fig. 3, "Walkway and ramp design") should be provided unless a safety harness is used in conjunction with the prearranged handles in way of the structure being accessed. Approval of portable platforms should address the following:

- Permissible loads,
- Permanent markings of the loads,
- Fixing arrangements,
- Guardrails; and
- Non-skid construction.

General Guidance

Safety measures should be taken by the authorised person prior to survey to the satisfaction of the attending surveyor(s). It should be confirmed that portable platforms are safely secured and supported prior to use. The maintenance of all equipment, the fixing of the equipment, its operation and training in its use should be addressed by the Ship's Safety Management System.

Scaffolding and Staging

Staging is the most common means of access provided especially where repairs or renewals are being carried out. Staging is generally an option for access to any structural members to be surveyed and measured in tanks, holds and spaces, but is **NOT** considered as an alternative to permanent means of access under the Technical Provisions Table 1—1.1.4 and Table 2—1.8. Staging not carried onboard is not subject to approval as part of SOLAS II-1/3-6. In this case, the asset owner and/or provider of the equipment are responsible for confirming safe use. Where staging and the associated equipment, including its attachments to the vessel's structure, are specifically designed for survey and thickness measurement in accordance with SOLAS II-1/3-6, such staging should be approved on behalf of the Flag State Administration based on a recognised international or national standard and necessary consideration is taken for the safety in the use.

Where staging is approved as a part of the Ship Structure Access Manual and carried onboard, the maintenance of all equipment the rigging of the equipment, its operation and training requirements in its use should be addressed by the Ship's Safety Management System.

General Guidance

The footing or anchorage for scaffolds should be sound, rigid and capable of carrying the maximum intended load without settling or displacement. Unstable objects such as barrels, boxes, loose brick or concrete blocks, should not be used to support scaffolds or planks. Scaffolds should have guardrails and toe boards installed on open sides and ends of the platform. A ladder or stairway should be provided for proper access and egress and should be affixed or built into the scaffold and so located that when in use it will not tip the scaffold. Riding on manually propelled scaffolds should not be allowed unless the following conditions exist:

- The floor or surface is within 3 degrees of level and free from pits, holes or obstructions,
- The minimum dimension of the scaffold base when ready for rolling is at least one-half of the height. Outriggers, if used, should be installed on both sides of staging,
- The wheels are equipped with rubber or similar resilient tyres; and
- All tools and materials are secured or removed from the platform before the mobile scaffold is moved.

Detailed Guidance

Scaffolds and their components should be capable of supporting at least 4 times the maximum intended load. When freestanding mobile scaffold towers are used, the height should not exceed four times the minimum base dimension. Casters are to be properly designed for strength and dimensions to support four times the maximum intended load. Casters are to be provided with a positive locking device to hold the scaffold in position.

Operational Considerations

Scaffolding is to be erected, moved, dismantled or altered under the supervision of qualified and trained personnel. Any scaffold including accessories (e.g., braces, brackets, trusses, screw legs, ladders, etc.) that are damaged or weakened from any cause should be immediately repaired or replaced. The force necessary to move the mobile scaffold should be applied as close to the base as practicable and provision should be made to stabilise the tower during movement from one location to another. Scaffolds should only be moved on floors that are level and free of obstructions and openings. Mobile scaffolds in use by any persons should rest upon a suitable footing and should stand plumb. The casters or wheels should be locked to prevent any movement. Slippery conditions on scaffolds should be eliminated as soon as possible after they occur.

Rafting

Rafting is generally used as a term for surveys carried out by means of boats or rafts. Rafting may be an option for use in tanks, holds and spaces which may be filled with water provided the arrangement of internal structure is as described in IACS Recommendation No. 39 "Safe Use of Rafts or Boats for Survey."

General Guidance

Surveys of tanks or spaces by means of rafts or boats may only be undertaken with the agreement of the attending surveyor(s), who is to consider the safety arrangements provided, including weather forecasting and ship response in reasonable sea conditions. The structural arrangement should allow easy escape to deck from any position being rafted. At least 1 m (39.0 in) clearance above and 0.5 m (20.0 in) clearance beyond the breadth of the raft should be allowed for the safe passage passed any internal obstructions. For bulk cargo holds designed for filling of water (e.g., ballast holds) and where filling up to a height not less than 2.0 m (79.0 in) below top of side frames is permitted (e.g., air draft holds), rafting may be utilised in lieu of permanent means of access to side frames provided the structural capacity of the hold is sufficient to withstand static loads at all levels of water needed to survey the side shell frames. Rafting of cargo tanks is subject to restrictions on discharging of water in harbour and weather conditions at voyage. Rafting as an alternative means of access should therefore not be considered as "readily accessible" in oil cargo tank and do not provide an alternative to fitting of longitudinal permanent means of access as required by Table 1—1.1.4.

Only rough duty, inflatable rafts or boats, having satisfactory residual buoyancy and stability even if one chamber is ruptured, should be used. The boat or raft should be tethered to the access ladder and an additional person should be stationed down the access ladder with a clear view of the boat or raft. Appropriate lifejackets should be available for all participants. The surface of water in the tank should be calm [under all foreseeable conditions the expected rise of water within the tank should not exceed 250 mm (10.0 inches)] and the water level stationary. On no account should the level of the water be rising while the boat or raft is in use. The tank or space must contain clean ballast water only. When a thin sheen of oil on the water is observed, further testing of the atmosphere is to be done to verify that the tank or space is safe for entering. At no time should the upside of the boat or raft be allowed to be within 1 m (39.0 in) of the deepest under deck web face flat so that the survey team is not isolated from a direct escape route to the tank hatch. Filling to levels above the deck transverses should only be contemplated if a deck access manhole is fitted and open in the bay being examined, so that an escape route for the survey party is available at all times. Rafts or boats alone may be allowed for close-up survey of the underdeck areas for tanks or spaces if the depth of the webs is 1.5 m (59.0 in) or less. If the depth of the webs is more than 1.5 m (59.0 in), rafts or boats alone may be allowed only:

- When the coating of the underdeck structure is in good condition and there is no evidence of wastage; or
- If a permanent means of access is provided in each bay to allow safe entry and exit. This means:
- Access direct from the deck via a vertical ladder and a small platform fitted approximately 2.0 m (79.0 in) below the deck in each bay; or

- Access to deck from a longitudinal permanent platform having ladders to deck in each end of the tank. The platform should, for the full length of the tank, be arranged in level with, or above, the maximum water level needed for rafting of under deck structure. For this purpose, the ullage corresponding to the maximum water level is to be assumed not more than 3 m (118 in) from the deck plate measured at the midspan of deck transverses and in the middle length of the tank. (See Figure below). A permanent means of access from the longitudinal permanent platform to the water level indicated above is to be fitted in each bay (e.g., permanent rungs on one of the deck webs inboard of the longitudinal permanent platform).

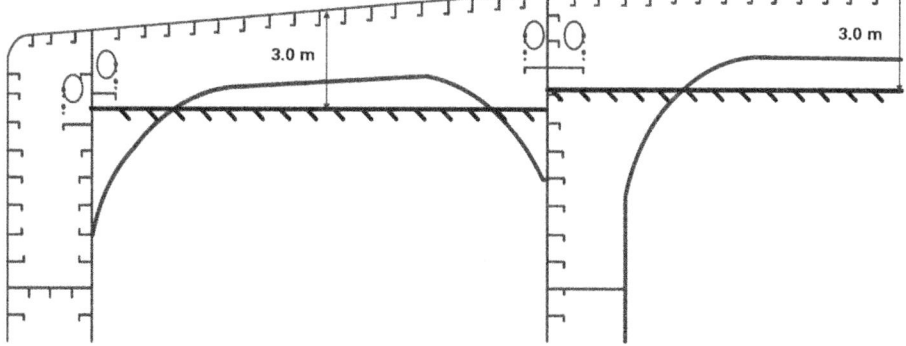

Operational Considerations

Rafting should be discontinued if the motion of the vessel (rolling) makes the operation difficult or hazardous. Factors such as the degree and period of roll, head space, and expected manoeuvring should be considered. If the tanks (or spaces) are connected by a common venting system, or inert gas system, the tank in which the boat or raft is to be used should be isolated to prevent a transfer of gas from other tanks (or spaces). Personnel assigned to conduct entry into confined spaces must have authorisation for such activity from a designated work authorisation supervisor. All personnel conducting rafting operations in cargo tanks may conduct such operations if the tank atmosphere is tested for oxygen content, flammable vapor, and concentrations of toxic contaminants and found to be safe for entry. At least one personal monitor (oxygen monitor or a combination oxygen/flammability/toxic meter) should be required when entering a confined space. An Emergency Escape Breathing Device (EEBD) should be required if there is a potential for a dynamic change in the environment such as a valve being opened and cargo entering the space, a bulkhead giving way and permitting entry of cargo, a space where inerted gas may be inadvertently introduced, etc. This equipment may not normally be needed if the space is tested prior to entry and ventilation is maintained. Adequate communication arrangements and equipment should be prepared for verifying the following:

- The attending surveyor(s) is always accompanied by at least one responsible person assigned by the company experienced in tank and enclosed spaces inspection. In addition, a backup team of at least two experienced persons should be stationed at the hatch opening of the tank or space that is being surveyed. The back-up team should continuously observe the work in the tank or space and should keep lifesaving and evacuation equipment ready for use,
- A communication system should be arranged between the survey party in the tank or space being examined, the responsible officer on deck, the navigation bridge and the personnel in charge of handling the ballast pump(s) in the pump control room. These communication arrangements should be maintained throughout the survey,
- Adequate and safe lighting should be provided for the safe and efficient conduct of the survey; and
- Adequate protective clothing should be made available and used (e.g., safety helmet, gloves, safety shoes, etc.) during the survey.

Remotely Operated Vehicle (ROV)

The interface between ROV system and support vessel/installation is defined as critical for efficient mobilisation and use of ROV systems. Typical interfaces/issues include:

- Weight of unit should be within the deck loading. The weight of all major components should be verified and registered prior to mobilisation,
- Sufficient power available and terminated in J-box,
- Connection points for communication, data transmission and video distribution,
- Easy and safe access between control station and launch location. Launch position should be free of obstructions and at a safe distance from vessel thruster propellers (if applicable),
- Available fresh water to be used for wash-down of the system,
- Required protected area for maintenance work when required; and
- System deck area should be kept tidy and free for hazards, and all hoses on deck should be secured and protected.

When the ROV system is intended to be operated in hazardous areas, it should be certified for such usage.

Appendix

American Bureau of Shipping. Guidance Notes for the Application of Ergonomics to Marine Systems. Houston, TX

American Bureau of Shipping. Guide for Crew Habitability on Ships. Houston, TX

American Society for Testing and Materials. (2000). Standard Practice for Human Engineering Design for Marine Systems, Equipment and Facilities (ASTM F 1166–2000). West Conshohocken, PA

International Association of Classification Societies, Recommendation No. 39—Safe Use of Rafts or Boats for Survey—IACS, Rev.3 (March 2009)

International Association of Classification Societies, Recommendation No. 42—Guidelines for Use of Remote Survey Techniques—IACS, Rev.1 (May 2004)

International Association of Classification Societies, Recommendation No. 72—Confined Space Safe Practice—IACS, Rev.2 (April 2007)

International Association of Classification Societies, Recommendation No. 76—Guidelines for Surveys, Assessment and Repair of Hull Structure—Bulk Carriers—IACS, Corr.1 (September 2007)

International Association of Classification Societies, Recommendation No. 78—Safe Use of Portable Ladders for Close-up Surveys—IACS, (September 2002)

International Association of Classification Societies, Recommendation No. 90—Ship Structure Access Manual—IACS, (October 2005)

International Association of Classification Societies, Recommendation No. 91—Guidance for Approval/Acceptance of Alternative Means of Access—IACS, Rev. 1 (January 2011)

International Association of Classification Societies, Unified Interpretations (UI) SC [191] for the application of amended SOLAS regulation II-1/3-6 (resolution MSC.151 (78)) and revised Technical provisions for means of access for inspections (resolution MSC.158 (78))

International Association of Classification Societies, Unified Requirements—Z10.1 Hull Surveys of Oil Tankers

International Association of Classification Societies, Unified Requirements—Z10.2 Hull Surveys of Bulk Carriers

International Association of Classification Societies, Unified Requirements—Z10.4 Hull Surveys of Double Hull Oil Tankers

International Association of Classification Societies, Unified Requirements—Z10.5 Hull Surveys of Double Skin Bulk Carriers

International Labour Organisation (1990). International data on anthropometry. Occupational Safety and Health Series: No. 65. Geneva

International Maritime Organisation, Maritime Safety Committee Resolution MSC.133(76) (adopted on 12 December 2002), Adoption of Amendments to the Technical Provisions for Means of Access for Inspections

International Maritime Organisation, Maritime Safety Committee Resolution MSC.134(76) (adopted on 12 December 2002), Adoption of Amendments to the International Convention for the Safety of Life at Sea

International Maritime Organisation. Maritime Safety Committee Resolution MSC.151(78) (adopted on 20 May 2004), Adoption of Amendments to the International Convention for the Safety of Life at Sea, 1974, as amended

International Maritime Organisation Maritime Safety Committee Resolution MSC. 158(78) (adopted 20 May 2004), Amendments to the Technical Provisions for Means of Access for Inspections

International Maritime Organisation. Maritime Safety Committee Resolution MSC.194(80) (adopted on 20 May 2005), Adoption of Amendments to the International Convention for the Safety of Life at Sea, 1974, as amended

International Maritime Organisation, SOLAS regulation II-1/3-6, "Access to and Within Spaces in, and Forward of, the Cargo Area of Oil Tankers and Bulk Carriers

International Organisation for Standardisation. (2008). Ship and marine technology—Identification colours for the contents of piping systems (ISO14726:2008), Geneva

International Organisation for Standardisation. (2001). Safety of machinery—Permanent means of access to machinery—Part 2: Working platforms and walkways (ISO14122-2:2001(E)) Geneva

International Organisation for Standardisation. (2001). Safety of machinery—Permanent means of access to machinery—Part 3: Stairs, stepladders and guardrails (ISO 14122-3:2001) Geneva

International Organisation for Standardisation. (2004). ISO 14122-4:2004—Safety of machinery—Permanent means of access to machinery—Part 4: Fixed ladders (ISO 14122-4:2004) Geneva

UK Department of Trade. (1998). Adultdata, the handbook of adult anthropometric and strength measurements—Data for design safety. Nottingham: Department of Trade

US Department of Labor (2000). Code of Federal Regulation, 29 CFR 1910. 23. Subpart D—Walking-Working Surfaces—Guarding floor and wall openings and holes. Washington, DC: Author

US Department of Labor (2000). Code of Federal Regulation, 29 CFR 1910. 24. Subpart D—Walking-Working Surfaces—Fixed industrial stairs. Washington, DC

Woodson, W.E., Tillman, B., and Tillman, P. (1992). Human factors design handbook: Information and guidelines for the design of systems, facilities, equipment and products for human use (2nd ed.). New York: McGraw-Hill, Inc

The manufacturer's authorised representative in the EU is Springer Nature Customer Service Centre GmbH, Europaplatz 3, 69115 Heidelberg, Germany. If you have any concerns regarding our products, please contact ProductSafety@springernature.com

Printed and bound by CPI Group (UK) Ltd, Croydon, CR0 4YY

26/03/2026

02078941-0015